SOLAR HYDROGEN:
Moving Beyond Fossil Fuels

Joan M. Ogden
Robert H. Williams

WORLD RESOURCES INSTITUTE
A Center for Policy Research

October 1989

Library of Congress Cataloging-in-Publication Data

Ogden, Joan M., 1950–
  Solar hydrogen.

  Includes bibliographical references.
  1. Solar energy.   2. Hydrogen as fuel.
I. Williams, Robert H., 1940– .   II. Title.
TJ810.O33   1989   333.792′3   89-22480
ISBN 0-915825-38-4

Each World Resources Institute Report represents a timely, scientific treatment of a subject of public concern. WRI takes responsibility for choosing the study topics and guaranteeing its authors and researchers freedom of inquiry. It also solicits and responds to the guidance of advisory panels and expert reviewers. Unless otherwise stated, however, all the interpretation and findings set forth in WRI publications are those of the authors.

Contents

Foreword

After a decade in hiding, the energy crisis is back. But it is wearing a new face. In the United States, total energy consumption is higher than it was when the oil shocks of the 1970s registered, and imports now account for roughly 40 percent of U.S. oil supply, compared with about 35 percent then. And today, we are also concerned about a complex and interrelated set of environmental issues.

For every mile driven by an automobile, about one pound of carbon dioxide is released into the atmosphere, where it adds to the greenhouse gas buildup that many scientists consider the biggest threat to the global environment and, ultimately, to the international economy. Recent forest and crop losses and the acidification of lakes also can be traced back to fossil fuel use. Some 110 million Americans live in areas that do not meet EPA air-quality standards, and President Bush and the City of Los Angeles recently have announced ambitious anti-pollution plans based partly on a switch to cleaner-burning automotive fuels.

Lacking the sudden grand entrance of the last energy crisis, the current crisis appears to be more entrenched, multi-dimensional, and long-lived. And because pollution is now just as big an issue as security of supply and oil prices, energy conservation can take us only so far. Sooner than most analysts imagined ten years ago, we will have to turn to alternative fuels; first for transportation.

Against this sobering backdrop, the prospect of hydrogen fuel produced with solar energy is exciting. From a strictly technical standpoint, hydrogen is almost an environmentalist's dream come true. It emits no carbon monoxide, volatile organic compounds, or particulates (troublesome urban air pollutants), no sulfur dioxide (a precursor of acid rain), and no carbon dioxide (the principal greenhouse gas). In fact, the only pollutant it emits is nitrogen oxide, which can be kept at safe levels quite easily. When hydrogen burns, the main byproduct is water vapor.

Hydrogen has more than clean air to recommend it. Produced via electrolysis from photovoltaic electricity using thin-film silicon solar cells, the hydrogen would not be resource-constrained. The main raw material used in the manufacture of these solar cells is silicon derived from sand, and the total land requirements for hydrogen production would be modest. Recent unanticipated advances in thin-film solar cell technology mean that solar hydrogen produced in the sunny American Southwest would be roughly competitive with synthetic fossil fuels by the turn of the century. While this hydrogen would not offer lower energy production costs, no billion-dollar capital investments such as those needed to launch synfuel or nuclear power plants would be required. And, as Joan Ogden and Robert Williams argue in *Solar Hydrogen: Moving Beyond Fossil Fuels,* if the environmental costs of *not* weaning ourselves from fossil fuels are taken into account, the balance would tip in favor of solar hydrogen.

Whether solar hydrogen gets the chance in our energy economy that it so richly deserves is, of course, a political decision. As Ogden and Williams note, fossil fuels enjoy so many direct and indirect subsidies that, without changes in public policy, hydrogen would make its debut on a playing field that is far from level. Only by discontinuing such subsidies to conventional energy sources as the depletion allowance to the oil industry and by levying new oil taxes and carbon taxes (graduated so that the fuels that release the most carbon in processing and combustion cost the most to buy) can we give hydrogen legs to stand on in the marketplace. Additional funds for research and demonstration projects are needed also to speed the introduction of hydrogen on the energy scene.

If the United States does not proceed seriously with the transition to hydrogen, other countries will. For example, West Germany has ambitious research and development programs for both solar hydrogen and hydrogen-powered automobiles.

Whether moved by the desire for cleaner air, worry about continued dependence on foreign oil supplies, or the spirit of economic competition, the United States faces fuel choices that will test the political system's capacity to anticipate trouble and to act before a full-fledged crisis is actually confronting the nation.

The World Resources Institute and the Center for Energy and Environmental Studies gratefully acknowledge support from the National Science Foundation, the Changing Horizons Charitable Trust, the Hewlett Foundation, the John D. and Catherine T. MacArthur Foundation, the Rockefeller Brothers Fund, and the Rockefeller Family and Associates for this study.

Mohamed T. El-Ashry, Ph.D.
*Vice President for Research
and Policy Affairs*
World Resources Institute

Acknowledgments

Many people helped shape our views, as our research on solar hydrogen evolved over the past three years. We are particularly indebted to David Carlson (Solarex), Henry Kelly (Office of Technology Assessment), and William Moomaw (World Resources Institute), who gave us extensive comments on several drafts of this book, and provided valuable insights on how we might best present our findings. This work has also benefitted from technical reviews by Rosina Bierbaum (Office of Technology Assessment), John Bockris (Texas A&M University), Bill Chandler (Battelle Laboratories), Napier Collyns (senior consultant, Cambridge Energy Research Associates), Al Mezzina (Brookhaven National Laboratory), Marc Ross (University of Michigan), Robert Socolow (Princeton University), Meyer Steinberg (Brookhaven National Laboratory), Patrick Takahashi (Hawaii Natural Energy Institute), Ted Taylor (consultant), Sigurd Wagner (Princeton University), Michael Walsh (consultant), Jerome Weingart (consultant), and Ken Zweibel (Solar Energy Research Institute). In addition, conversations with Roger Billings (American Academy of Science), David Block (Florida Solar Energy Center), Mark DeLuchi (University of California, Davis), Peter Hoffmann (The Hydrogen Letter), Sandy Kaplan (Chronar), Philip Metz (Brookhaven National Laboratory), Jochim Nitsch (DFVLR, Federal Republic of Germany), Jack Pfister (Salt River Project, Arizona), Thomas Schucan (Swiss Federal Institute for Reactor Research), Z Smith (Xerox), Daniel Sperling (University of California, Davis), Harmut Steeb (DFVLR, Federal Republic of Germany), and Carl-Jochen Winter (DFVLR, Federal Republic of Germany) provided useful insights during the course of this work.

We also thank David Sheridan and Kathleen Courrier for their editorial efforts in bringing the manuscript to its final form and Hyacinth Billings for production supervision.

Support for this research was provided by the National Science Foundation, the Hewlett Foundation, the Changing Horizons Charitable Trust, the Rockefeller Brothers Fund, the Rockefeller Family and Associates, and the World Resources Institute.

J.M.O
R.H.W.

I. Introduction

The prospect of using hydrogen from water as a substitute for oil and other fossil fuels has moved, with very little notice by energy planners and decisionmakers, from the realm of dream to distinct possibility.

How to split water into hydrogen and oxygen by passing an electric current through it has been known since the early 1800s. And industry has used this process, electrolysis, for many years to produce hydrogen as a chemical feedstock. But the great stumbling block to electrolytic hydrogen as a competitive fuel has been the cost of electricity.

Back in the heady early days of nuclear power, it was thought that nuclear-generated electricity would be "too cheap to meter," so that hydrogen derived from this electricity would be able to supplement and eventually supplant oil and other fossil fuels. However, low-cost electricity from nuclear power has turned out to be an elusive goal.

What has made hydrogen worth reconsidering now are recent dramatic developments in photovoltaic (PV) cells that directly convert sunlight into electricity. These advances (specifically, new technologies that can be mass produced at much lower costs than the solar cells that were the main focus of photovoltaic R&D efforts in the 1970s) mean that in the sunnier regions of the United States, the direct current (DC) electricity needed to run electrolyzers to produce hydrogen could be available by the turn of the century at $0.020 to $0.035 per kilowatt-hour (kWh). Hydrogen produced from PV electricity at this cost would be sufficiently economical to contribute substantially to society's efforts to cope with the pressing fossil fuel-related problems of urban air pollution, regional acid precipitation, and global warming.

Hydrogen is an exceptionally clean-burning fuel—cleaner than today's fossil fuels and such proposed fossil fuel-based synfuels as methanol made from coal or natural gas. With hydrogen fuel there are no emissions of carbon monoxide, volatile organic compounds or particulates that pollute the air in so many urban areas, and no emissions of sulfur dioxide—one of the major causes of acid precipitation. Indeed, the only pollutant is nitrogen oxide, and it can be controlled to very low levels. When hydrogen is made from non-fossil sources, such as solar PV power, no carbon dioxide is emitted in the production or use of fuel. Indeed, PV hydrogen is one of the few long-term energy supply options that could meet the world's energy needs without contributing to the greenhouse effect. PV hydrogen thus has a decided edge over such synfuels as methanol, which are currently being considered as alternatives to oil. Strictly in terms of fuel production costs, PV hydrogen might not be preferable to methanol derived from fossil fuels, but if the major environmental costs of the two fuels are included in the comparison, then PV hydrogen is clearly preferable.

1

Because it is based on the exploitation of renewable resources (water and sunlight) and on abundant materials (silicon from ordinary sand in the case of amorphous silicon thin-film cells), PV hydrogen is a fuel with none of the resource limitation worries of oil, natural gas, or uranium. Moreover, it would be feasible to meet the entire fuel requirements of the United States with PV hydrogen produced domestically without running up against serious land-use constraints. Indeed, relatively low-value arid lands in sunny areas can be used for the solar collector fields needed to produce PV hydrogen.

Another noteworthy advantage of PV hydrogen is that it does not require huge, billion-dollar capital investments, like fossil fuel-based synfuel or nuclear power plants do, to attain economies of scale in production. In fact, it appears that there are no economies of scale to be achieved in PV hydrogen production beyond a relatively modestly sized 5 to 10 megawatt facility which would cost only $4 million to $12 million to build.

How and when might a transition to a PV hydrogen economy begin in the United States? As we shall show, a hydrogen economy stands the best chances of succeeding if hydrogen serves energy-efficient end-use technologies. Accordingly, a precondition for a successful transition to a hydrogen economy appears to be emphasis in the near term on reducing fossil fuel use by improving energy efficiency.

The transition to hydrogen as an energy carrier could begin with small local energy systems located in the sunny Southwest, where the cost of delivered PV hydrogen would be the most favorable. The modular nature of PV hydrogen technology makes it well-suited to low-cost, near-term demonstrations (important, given the mounting political pressure to do something about the greenhouse effect). Initial applications of hydrogen as an energy carrier will probably be in the urban transport sector since transportation fuels are the most expensive fossil fuels, the prospects are good that

hydrogen-powered cars could be commercialized over the next decade (especially in Germany and Japan where development is underway), and there are severe and growing air-pollution problems associated with automotive emissions in several urban areas of the Southwest.

Initially, perhaps within the next five to ten years, fleet vehicles in some Southwestern city could be converted to hydrogen produced off-peak at some low-cost, existing power plants in the region, to gain experience with hydrogen fuel while the costs of PV hydrogen come down. If this demonstration effort were to succeed, then by the turn of the century PV hydrogen systems could provide additional hydrogen supplies, as the cost of PV hydrogen becomes economically attractive.

Once the technology is established for fleets, the local utility could begin to offer hydrogen fuel for private automobiles. To interest consumers, it would be important to have multi-fuel capable vehicles, which could use other fuels in areas where hydrogen would not yet be available.

If local hydrogen transport systems come to be viewed as an attractive way to cope with the problem of urban air pollution in the Southwest, cities in the Northeast and the Midwest might decide to convert to hydrogen as well. If 10 percent of fleet vehicles in the United States converted to hydrogen, this would be enough demand to justify building a pipeline to bring hydrogen from the Southwest to the Northeast. Hydrogen-powered transport would then follow a similar pattern in northern cities, serving first as a fuel for fleets and then for private vehicles.

Once pipelines are built and hydrogen is established as a transportation fuel, other uses might be found as well. For example, hydrogen could be used for residential space and water heating in the Northeast.

A switch to hydrogen fuel would take many decades to complete, but it seems that significant initial benefits from hydrogen as a low-polluting alternative to fossil fuels could be realized beginning around the turn of the century.

It is, we think, a plausible scenario. But the pace at which it will unfold depends upon the extent to which our society reduces its dependence on fossil fuels as a response to the worsening crises of urban, regional, and global pollution problems their production and use causes. Individual consumers acting in their own economic self-interest will *not* forestall the global climate disruptions of the greenhouse effect; they will *not* reduce urban air pollution to levels acceptable for human health; they will *not* prevent further acid precipitation damage to lakes, crops and forests. Public policy initiatives will be needed to facilitate a shift from fossil fuels to PV hydrogen.

II. Needed: A Low-Polluting Alternative to Fossil Fuels

The Emerging Fossil Fuel Crisis

The oil crises of the 1970s are fading into history. While their root cause remains—the concentration of remaining global oil resources in the Middle East—the world oil market is no longer being manipulated by Middle Eastern and other OPEC oil producers and will probably not be under their control again for at least a few more years, owing to successful oil-conservation efforts by industrial countries in the late 1970s and early 1980s and increased oil production by non-OPEC producers.[1]

But a new energy crisis is rapidly unfolding—a crisis that may make the oil price shocks of the 1970s seem like minor tremors in our energy system. The emerging crisis relates not to issues of energy supply but to a complex web of environmental problems caused by the use of fossil fuels that not only endangers the quality of life in modern society but also jeopardizes continued global development. Deteriorating urban air quality, crop losses, the acidification of lakes and forests, and a changing global climate are interrelated consequences of fossil fuel use that are making fossil fuel pollution front page news and are prompting policy-makers to propose far-reaching new environmental control policies.

Urban Air Pollution

By far the most familiar and easy-to-grasp air pollution problem is urban air pollution. It can often be seen and smelled, and its adverse effects on human health and property are fairly well understood. Since the passage of the Clean Air Act in 1970, ambient levels of airborne lead and carbon monoxide have been reduced, but the urban air pollution problem is far from solved; in 1987, over 130 million Americans lived in areas where ozone levels in photochemical smog exceeded federal air quality standards.[2] Exposure to levels of ozone in excess of the present federal standard (0.12 parts per million) can cause acute respiratory symptoms and has been implicated in the onset of respiratory disease.[3] Recent studies have raised concern that chronic exposures to lower ozone levels (in the range 0.08 to 0.12 parts per million) may lead to permanent lung damage.[4]

When the Clean Air Act was passed in 1970, the final deadline for compliance with ozone standards was set for 1975. In 1977, with 78 areas still out of compliance, the deadline was pushed ahead to 1982. By 1983, ozone standards had been relaxed from 0.08 ppm to 0.12 ppm, but many areas were unable to meet even these less stringent standards and obtained deadline extensions to December 31, 1987. Twenty days before the end of 1987, the deadline was extended again to August 31, 1988.[5] Early in 1988, the Environmental Protection Agency (EPA) projected that some 68 cities would again fail to meet the extended deadline.[6] The summer of 1988 brought a stifling combination of heat, sun, and stagnant weather patterns—which some scientists see as

5

early symptoms of the greenhouse effect[7]—that drove ozone pollution in many U.S. cities to the highest levels this decade.[8] It now appears that many more than 68 cities were out of compliance in 1988.

Ozone is formed when nitrogen oxides (NO_x) and volatile organic compounds (VOCs) react in the presence of sunlight. NO_x is a product of combustion at high flame temperatures in power plants and motor vehicles. VOCs are emitted as unburned hydrocarbons in car exhaust, gasoline pump vapors, and fumes from solvents in paints and other chemicals. Reducing ozone is particularly difficult because such a wide variety of NO_x and VOCs sources must be controlled.

Motor vehicles are large sources of both NO_x and VOCs. In 1985, motor vehicles contributed an estimated 45 percent of the NO_x and 34 percent of the VOCs emitted in the U.S. In some cities the percentages from mobile sources were even higher.[9] But since 1970, progressively stricter emissions standards for vehicles have been implemented, and, for automobiles, the actual NO_x and VOCs emissions per-mile-travelled decreased fourfold between 1972 and 1988.[10] Yet, due to the growing number of autos, the lack of stringent standards for heavier vehicles, and the increased number of vehicle miles travelled, aggregate NO_x emissions from all vehicles have stayed essentially constant and VOCs emissions have decreased only slightly. (See Figure 1.) Calculations carried out by the EPA indicate that both VOCs and NO_x levels could increase sharply over the next few decades if emissions from both mobile and stationary sources are not reduced. (See Figure 2.)

Control of emissions from motor vehicles must be an important part of an overall ozone-control strategy. In the near term, much could be accomplished by instituting tougher vehicle-emissions standards. If proposed emissions standards for motor vehicles advanced in recent federal legislation (S. 1894, adopted by the Senate Environment and Public Works Committee in 1987 and passed by Congress in 1988) were met, total U.S. emissions of NO_x and VOCs from all sources could be stabilized or even decreased slightly over the next couple of decades.[11]

Meeting the proposed standards for automobiles would require a threefold reduction in the average actual in-use emissions. (See Table 1.) While major changes in current practice would be required to meet the new standards, doing so would be technically feasible. The standards could be met in part by improving maintenance and rigorously enforcing inspection programs to make vehicles in routine use operate much closer to the levels certified for new cars. Technical improvements could also reduce emissions per mile.

But tighter emissions controls alone will not be enough to bring all metropolitan regions into compliance with the ozone standards. Even with tough controls on both mobile and stationary pollution sources, the Los Angeles-Long Beach, California metropolitan area would be unable to comply with the ambient air quality standards for ozone.[12] From 1983 through 1985, the so-called design value (a measure based on the fourth highest of all the daily peak 1-hour average ozone concentrations in the previous three years) was 0.36 ppm, or three times the federal standard. And other rapidly growing metropolitan areas (Phoenix, for instance) would be in compliance for at most a few years before the burgeoning vehicular population would once again convert these regions into nonattainment areas.

The limitations of tighter emissions controls in combatting urban air pollution have prompted interest in cleaner transport fuels. The Alternative Fuels Act, passed by Congress in 1988, mandates the use of "alternative" motor fuel in some ozone nonattainment areas. This new law gives car manufacturers a powerful incentive to produce vehicles capable of using alternative fuels by significantly relaxing the fuel economy standards for such vehicles. For vehicles that use gasoline-alcohol blends, the law's fuel economy standard requires that

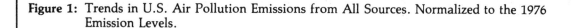

Figure 1: Trends in U.S. Air Pollution Emissions from All Sources. Normalized to the 1976 Emission Levels.

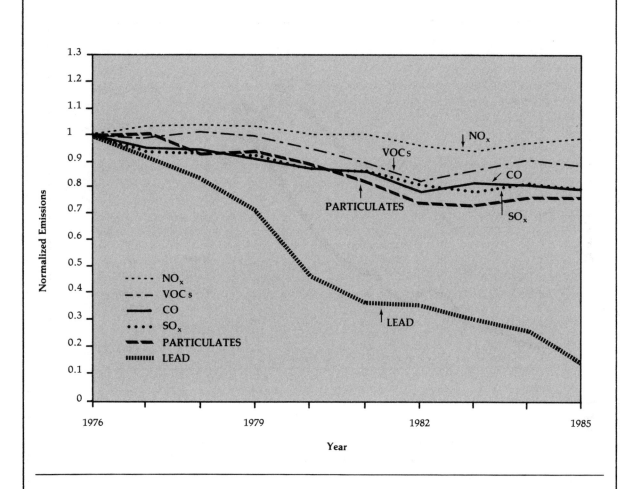

Source: Michael P. Walsh, "Pollution on Wheels," Report to the American Lung Association, February 11, 1988.

only the conventional component (gasoline) must be counted in determining compliance.

Perhaps the most ambitious alternative fuels plan advanced to date is the long-range plan announced June 30, 1988, by the California South Coast Air Quality Management District to bring the greater Los Angeles area into compliance with the federal standards for ozone.[13]

This plan, which is subject to EPA approval, proposes phasing out petroleum fuels in favor of alternative fuels in both industry and transportation. In the transport sector, the plan calls for converting 40 percent of passenger cars and 70 percent of trucks to alternative fuels.

The alternative fuel most often considered for reducing ozone is methanol (either 100 percent

Figure 2: Trends in NO_X and VOCs Emissions from All Sources in the United States. Historical Data and EPA Projections are Shown.

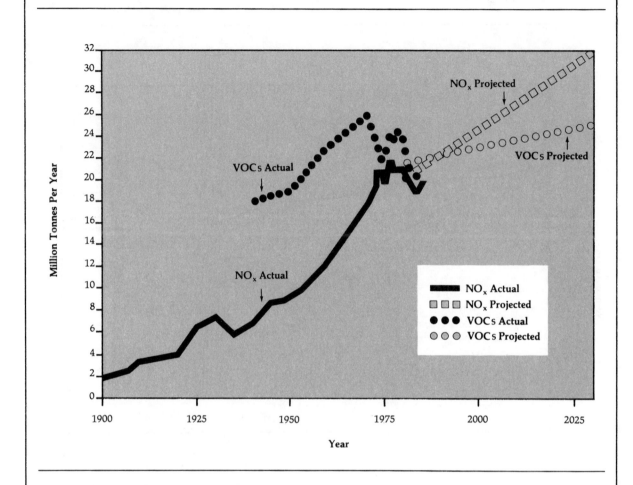

Source: Michael P. Walsh, "Pollution on Wheels," Report to the American Lung Association, February 11, 1988.

"neat" methanol or a blend of at least 85 percent methanol and up to 15 percent gasoline). From an air quality perspective, a major attraction of methanol is that it reacts much more slowly in the atmosphere and consequently produces less ozone than the VOCs emitted from the combustion and evaporation of gasoline. A recent EPA study estimates that, per mile travelled, substituting methanol for gasoline is 30 percent to 90 percent as effective in reducing ozone concentrations as completely eliminating emissions from gasoline-fueled vehicles.[14] Unfortunately, there is little on-the-road evidence to support these claims, and some studies suggest that gains are more modest.

While emphasizing methanol as a transport fuel might help improve urban air quality, a shift to methanol could significantly exacerbate another important environmental problem: the global greenhouse warming associated with the

Table 1. Automotive Emissions[a,b] (in grams/mile)

| | AVERAGE EMISSIONS | | EMISSIONS STANDARDS | |
	Certified (1987)	Actual (1986)	1988	Proposed
Automobiles				
Carbon Monoxide	1.91	10.0	3.4	3.4
Hydrocarbons (VOCs)	0.20	0.8	0.41	0.25
Nitrogen Oxides	0.37	1.2	1.0	0.4
Particulates	–	–	0.2	0.08

a. *Source:* Michael P. Walsh, "Pollution on Wheels," Report to the American Lung Association, February 11, 1988.

b. Shown here are the average certified emissions from 1987 gasoline-powered automobiles, the actual in-use emissions for 1986 cars, the 1988 standards for automotive emissions, and proposed standards. The certified emissions level is the average for new cars as measured for an idealized standardized driving cycle. The actual in-use level is the measured value for cars in the field under real operating conditions. The actual in-use emissions from 1986 cars are several times higher than the certified emissions, because cars are not operated precisely as modeled in the standardized driving cycles used for testing purposes and because few cars are maintained in optimum condition.

buildup in the atmosphere of carbon dioxide from the burning of fossil fuels. Initially, methanol for transport would be derived from natural gas, and carbon dioxide emissions would be comparable to those arising from gasoline. But because global natural gas resources are probably not much more abundant than global oil resources,[15] a "methanol economy" would, within decades, have to shift its base from natural gas to abundant low-quality feedstocks such as coal. In a transport sector operated on methanol derived from coal, carbon dioxide emissions per unit of useful energy derived would be roughly twice as large as for gasoline. (*See Figure 3.*)

Regional Air Pollution

In the 1960s and early 1970s, concerns about deteriorating air quality were focussed largely on local air pollutants arising from local sources. In this period many polluting utilities built tall stacks for their fossil fuel power plants to disperse the offending pollutants over a wide area. But it is now recognized that "dilution is not the solution to pollution," because certain dispersed pollutants cause major damage over wide areas. Indeed, the transnational transport of air pollutants has become a matter of international dispute.

The most-discussed form of regional pollution involves the deposition on lakes, farmland, and forests of acidic air pollutants, formed largely from emissions of nitrogen oxides and sulfur oxides from the combustion of fossil fuels in road transport and stationary power plants. This acid deposition (popularly known as acid rain even though the acids can also fall in snow, fog and dry forms) has raised the acidity of some lakes in the northeastern United States and southeastern Canada. Declines in fish and insect populations and loss of species diversity, which correlate directly with acidification, have been observed in these lakes.[16]

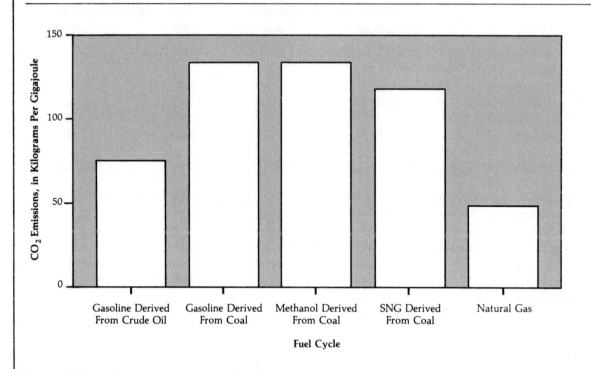

Figure 3: CO_2 Emissions from the Processing and Use of Various Fuels, in Kilograms of CO_2 Released Per Gigajoule (GJ) of Fuel Energy Used.

Here it is assumed that crude oil is refined to give gasoline and Diesel fuel with 10 % refinery losses; gasoline (plus liquid propane gas and synthetic natural gas as byproducts) is derived from coal via the SRC-II process; methanol is derived from coal via the Texaco gasifier process; synthetic natural gas is derived from coal via the Lurgi dry ash process.

Acid deposition combined with other air pollutants (such as ozone or heavy metals) and climatic stresses (such as the drought, hot summers, and cold winters of the past few years) may be a factor in the observed decline of forests in the northeastern United States and Canada.[17,18] In the United States, the red spruce forests on the upper slopes of the Appalachians, where the acid deposition is greatest, have been declining since the 1960s. On the highest mountains, the red spruce forests have already died, and some scientists fear for the lower elevation hardwood forests. To paraphrase one scientist, the Appalachian high forests may be like the coal-miner's canary.[19]

Ozone pollution can not only harm human health, the crux of concern regarding urban air pollution; it can also damage crops and forests. Concentrations as low as 0.04 to 0.06 parts per million, which are typical mean levels in many rural areas in the eastern United States, can seriously reduce crop yields.[20] Ozone may also be a factor contributing to the decline of forests in the United States, Canada, and Europe.[21]

While considerable uncertainties still surround the extent of damage caused by acid deposition and other long-range pollutants, a formidable body of scientific evidence already shows that the long-range pollutants caused by the burning of fossil fuels, often acting in concert with climatic stresses (that may also be related to the burning of fossil fuels, through the greenhouse effect), are significantly altering the ecology of lakes and reducing the productivity of farmland and forests.[22]

Global Air Pollution: The Greenhouse Effect

The combustion of fossil fuels leads to the release of carbon dioxide and its accumulation in the atmosphere. While carbon dioxide is a colorless, odorless gas that causes no direct damage to human health nor to the local or regional ecology, its buildup in the atmosphere can change the global climate through the greenhouse effect. Global warming, changes in precipitation patterns throughout the world, and a rise in sea level are some of the more serious repercussions of the greenhouse effect.[23]

Though uncertainties abound regarding the timing, severity, and the distribution of the consequences of the greenhouse effect, a scientific consensus has developed regarding the seriousness of the problem if the buildup of carbon dioxide in the atmosphere continues. Scientists' concern has been reinforced in the public's mind by the record temperatures (and associated drought, forest fires, and heightened air pollution) and intense hurricane activity experienced during the summer and fall of 1988. While it is premature to blame the greenhouse effect for these events, which may well be merely natural variations in the climate, experts do see these events as a foretaste of what would happen in the United States as the greenhouse effect worsens.[24]

Given this consensus, what do the large increases in fossil fuel consumption predicted for the next several decades[25] imply for future carbon dioxide levels in the atmosphere? Consider the 1981 global energy study of the International Institute for Applied Systems Analysis (IIASA). Typical of many global energy studies, the IIASA study projected that fossil fuel use would increase between 1.8-fold and 3-fold between 1980 and 2030.[26] In IIASA's scenarios,

the carbon dioxide level would increase from 1.25 times the pre-industrial level in 1984 to 1.55 to 1.65 times the pre-industrial level by 2030.[27] Climatic modelers believe that doubling the atmospheric carbon dioxide level from the pre-industrial level will increase the average global surface temperature by 1.5 to 4.5°C;[28] thus, under the IIASA scenarios the equilibrium average surface heating resulting from carbon dioxide released to the atmosphere by 2030 would be 60 to 70 percent of the heating arising with a doubling of atmospheric carbon dioxide.[29] The actual surface heating would probably be about twice as large, owing to the simultaneous buildup of other greenhouse gases (especially chlorofluorocarbons, ozone, methane, and nitrous oxide) associated with human activities.[30] In just a few decades time, human activities could thus lead to an eventual increase in the global temperature of 2°C or more, which would make the average surface temperature hotter than it probably has been during the last 850,000 years, with far-reaching implications for a wide range of human activities and the global environment.[31]

Although energy is a necessary input to all human activity, the extent of the greenhouse warming need not be as great as is indicated by the IIASA and related projections. Because the economic base of the industrialized countries is shifting toward less energy-intensive economic activities, energy demand in these countries (which account for half of the increment in global energy demand projected in the IIASA study) will probably grow much less rapidly than IIASA projected.[32] There are also many cost-effective opportunities for using energy more efficiently. Taking into account both ongoing structural changes in industrialized countries and cost-effective opportunities for using energy more efficiently in all countries, it has been shown that global economic goals could be met over the next several decades with little increase in global energy use.[33] For developing countries, emphasis on energy efficiency is not just an option but a necessary condition for meeting development goals. In general, it requires much less capital to improve energy efficiency than to expand energy supplies—a crucial consideration given the severe capital constraints most developing countries face today.[34]

While improved energy efficiency could stabilize global fossil fuel use over the next several decades, this strategy alone would not solve the greenhouse problem. Even with global fossil use held constant, the atmospheric carbon dioxide level in the atmosphere by 2020 would still be some 40 percent higher than the pre-industrial level, and the eventual global warming from carbon dioxide in the atmosphere at that time would still be about half as much as with a doubling of the carbon dioxide level.[35]

However difficult to accept, it may be both necessary and desirable *to reduce fossil fuel use globally by 50 to 75 percent or more in the coming decades,* so as to stabilize the atmospheric level of carbon dioxide and limit the adverse consequences of the greenhouse warming. The urgency of doing so was reflected in Toronto in 1988 at the World Conference on The Changing Atmosphere. The final formal policy statement of this historic conference, hosted by the Canadian government with the support of the United Nations Environment Programme and the World Meteorological Organization, set, as a first step, a target of reducing annual global carbon dioxide emissions 20 percent by the year 2005 through improved energy efficiency and shifts in the energy supply mix.[36] The message is clear: improved energy efficiency is one part of a solution to the greenhouse problem, and a global shift to fuels that emit less greenhouse gases is another.

The Hydrogen Economy Revisited

Hydrogen as an Alternative to Fossil Fuels

Hydrogen is an extraordinarily clean, high-quality fuel that could be used instead of oil and natural gas for transportation, heating, and power. When hydrogen is burned, the primary combustion product is water vapor. No carbon monoxide, carbon dioxide, sulfur dioxide, VOCs, or particulate matter is produced in combustion. When hydrogen is burned in air at high temperatures, however, nitrogen oxides are formed. But when only one pollutant is formed in combustion, its control is far easier than would be the case where multiple pollutants must be controlled simultaneously—for example, in a gasoline- or diesel-fired internal combustion engine. Hence NO_x emissions can with various techniques be kept to very low levels in hydrogen-fueled engines. With hydrogen fuel cells that may one day be practical commercial devices, even uncontrolled NO_x emissions would be negligible.

Much of the hydrogen used in industry today is derived from natural gas. Of course, using hydrogen derived from fossil fuel would not help society cope with the greenhouse problem. If a fossil fuel is the feedstock, carbon dioxide would be released to the atmosphere during production, though not during combustion. But hydrogen can also be made without using fossil fuels—for instance, through the electrolysis of water using electricity from solar, wind, hydropower or nuclear sources.

Given this environmental advantage, a large-scale hydrogen energy system or "hydrogen economy" based on non-fossil fuel feedstocks has long been seen as ideal for meeting global energy needs.[37] But, thus far, a hydrogen economy has remained an elusive dream, because no source of electricity has yet been developed that is inexpensive and widely available enough to enable electrolytic hydrogen to compete economically with carbon-based fuels.

The prospects for a nuclear power-based hydrogen economy looked bright in the 1960s, when it was thought that nuclear power would soon become "too cheap to meter." But the promise of low-cost nuclear power has not materialized. In fact, nuclear power has priced itself out of the electric power market in the United States, and, worldwide, nuclear power is the focus of much public concern and anxiety as a result of the accidents at Three Mile Island and Chernobyl, the unsolved problems of nuclear waste disposal, and the possibilities for the proliferation of nuclear weapons-usable materials diverted from nuclear power fuel cycles. Moreover, even if the hoped-for improved economics of a born-again nuclear industry could be realized, nuclear electricity would still not be cheap enough to make the derived hydrogen an economically attractive alternative to carbon-based fuels. *(See Chapter 5.)*

What about renewable sources of electricity—hydroelectric, wind, and photovoltaic sources? Low-cost hydrogen can be produced at many existing hydroelectric sites, but global hydroelectric resources are not plentiful enough to allow hydrogen produced this way to have much of a global impact in replacing fossil fuels,[38] and new hydroelectric supplies are generally much more costly than the low-cost supplies already developed.[39] Wind energy resources are similarly limited in global extent,[40] and wind-based hydrogen would also be relatively costly. *(See Chapter 5.)*

In contrast, the photovoltaic (PV) resource is much less geographically limited than hydro- and wind-power: PV hydrogen can be produced in any sunny region. *(See Figure 4.)* In principle, PV hydrogen can be produced in quantities sufficient to allow it to substitute for fossil fuel on a global scale.

PV hydrogen systems were studied intensively in the 1970s. Research indicated that it would be technically feasible to produce PV hydrogen and use it for transport, heating, and power. However, economic assessments published in the early 1980s concluded that solar cells, and therefore PV electricity, would probably always be too expensive for PV hydrogen to compete with other fuels. This conclusion, based largely on late 1970s projections of the eventual cost and efficiency of solar cells, has remained the "conventional wisdom," at least in the United States.

The Prospects for PV Hydrogen Derived Using Thin-Film Solar Cells

While relatively little attention outside of West Germany has been paid to PV hydrogen since the early 1980s, there has been a revolution in the development of new, inexpensive solar cell materials. Particularly rapid progress is being made with thin-film solar cells. One promising thin-film technology is the amorphous silicon (a-Si) solar cell—a technology that even theoretical physicists had not imagined before the mid–1970s. Over the last few years, increases in the efficiency and reductions in the costs of thin-film solar cells have been dramatic and are expected to continue. It now appears that solar cells and PV electricity may turn out to be much less expensive than was thought possible just a few years ago. *(See Chapter 3.)*

Solar cells and PV electricity may turn out to be much less expensive than was thought possible just a few years ago.

Ongoing and projected advances in thin-film solar cells have profound implications for PV hydrogen. To indicate the possibilities, we show here that if the PV industry's projections for a-Si solar cells are achieved, PV hydrogen could become roughly cost-competitive with synthetic fuels derived from coal or other sources by the early part of the next century. *(See Chapter 5.)* By then, the consumer would probably pay about as much to use low-polluting PV hydrogen as to use coal-based synthetic fuels or electricity from coal or nuclear sources. *(See Chapters 6 and 7.)* Considering its environmental benefits relative to fossil fuels—clean air and no greenhouse gases—and the opportunity to gain these benefits without resorting to large-scale nuclear power development with its attendant risks, PV hydrogen looks to be the more attractive option.

13

Figure 4: A Solar Photovoltaic Electrolytic Hydrogen System.

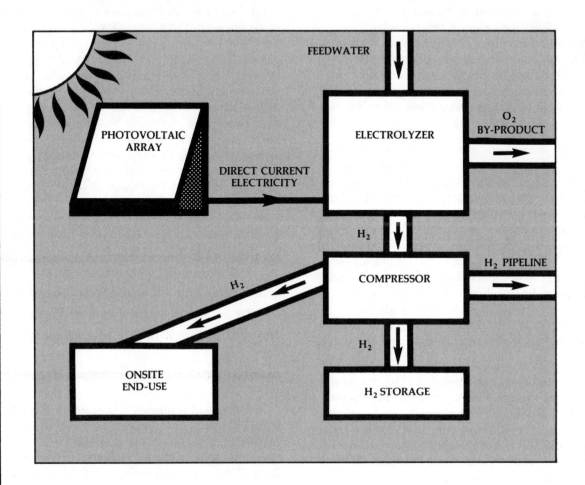

A PV array converts sunlight into DC electricity, which powers an electrolyzer, splitting water into its constituent elements, hydrogen and oxygen. A compressor pressurizes the hydrogen for storage, onsite use or pipeline transport to distant markets.

This outlook for solar hydrogen, which contradicts the accepted view, has eluded energy planners for several reasons. First, the prospect of dramatic advances in thin-film solar cell technology has become clear only recently. For example, in the early 1980s, when the last major assessments of PV hydrogen were completed, amorphous silicon solar cells were still only laboratory curiosities, and it was thought that commercial PV power would require crystalline devices, which are inherently more costly than amorphous silicon devices. Second, low world oil prices have tended to keep energy issues off the public policy agenda during the last few years, so that policy-makers and the public generally are unaware of the spectacular progress being made in the PV industry. And, third, even within the PV research and development community, efforts have been focussed on the nearer-term PV electricity markets rather than on longer-term PV hydrogen markets.

III. Photovoltaic Power Comes of Age

The prospect that low-polluting PV hydrogen could begin to be used as a transport fuel by the turn of the century is not rooted in any dramatic developments relating to hydrogen. Rather, it is a direct result of rapid advances being made in PV technology.

Compared to other methods of electric generation, using PV cells to convert sunlight into electricity is exceptionally clean and quiet. The process entails no moving parts and no pollution, the "fuel" is free and inexhaustible sunlight, and many PV technologies involve the use of only such plentiful raw materials as silicon, which can be obtained from ordinary sand.

Until recently, the solar cell was considered an "exotic" power source—too expensive for all but a few specialized applications such as spacecraft, navigation buoys, and remote weather stations. In the last few years, the development of new solar cell materials has radically changed this picture. These new materials require much less raw material and energy to manufacture, and they can be mass produced inexpensively.

Before about 1980, the only commercially available solar cells were made of high-grade single-crystal silicon. Growing crystals for these cells is time-consuming and requires large amounts of material and energy.[41] In the early 1970s, crystalline silicon solar cells sold for about $120 per peak Watt.[42,43] Such high costs were tolerable for space power, but they implied that terrestrial power applications were but dreams. The cost of electricity from solar cells priced at $120 per peak Watt would be more than $5 per kilowatt-hour (kWh),[44,45] compared to electricity from new coal and nuclear power plants costing $0.05 to $0.06 per kWh.[46] Improved manufacturing techniques have enabled manufacturers to cut the prices of crystalline solar cells to about $5 per peak Watt ($5/Wp), but this price is still too high for large-scale use.

Further cost reductions, as well as increases in efficiency to 30 percent or more, will probably be achieved with further development of crystalline silicon cells.[47] Producing electricity competitively with single-crystal solar cells may well require the use of concentrating solar collectors, which can help make these relatively high-cost cells more competitive but which also add to the mechanical complexity of the system.[48]

Various thin-film PV technologies probably offer the best prospects for lowering capital costs, though the efficiencies that can be realized with potentially low-cost, thin-film devices are lower than what is possible with crystalline cells. High on the list of promising thin-film technologies are amorphous silicon (a-Si) cells[49] and polycrystalline copper indium diselenide ($CuInSe_2$) and cadmium telluride (CdTe) cells—all of which offer the advantages of low materials requirements and cost-cutting through the use of automated manufacturing processes.[50]

The analysis here is focussed on a-Si thin-film PV technology. There are several reasons for this choice. First, as a thin-film technology under rapid development, the a-Si option offers good prospects for substantial cost reduction soon. Second, no obvious resource or environmental constraints limit the large-scale development of a-Si technology. And, third, the a-Si manufacturing process is sufficiently well understood that a plausible quantitative analysis can be presented showing how costs might be lowered as the technology advances. Of course, this last reason does not represent an inherent advantage of a-Si technology—it merely makes possible a fairly solid analytical basis for estimating future cost trends.

Though we have focussed on a-Si PV technology for specificity's sake, it is premature to declare a-Si technology the winner of the PV race. The case made here would hold if one of the other leading contenders were to win this fiercely competitive race.

The Amorphous Silicon Solar Cell

In the mid–1970s, it was discovered that thin films of amorphous silicon could be used to convert sunlight directly into electricity. Only 1 micron thick, these thin films use much less raw material than crystalline solar cells, which are typically 100 to 200 microns thick. In fact, so little material is required for the active layer of a-Si cells that the amount of electricity that can be extracted from a gram of silicon over a solar cell's life is comparable to the amount that could be extracted from a gram of uranium used in a plutonium fast breeder reactor! Of course, silicon is also far more abundant than uranium in the earth's crust. (See Box 1.)

Another attractive feature of a-Si cells is manufacturing ease. Instead of the painstaking crystal growing-and-cutting process, production of a-Si solar cells involves a simpler process of depositing silicon vapor on an inexpensive substrate such as glass, plastic, or stainless steel. Because of the speed with which vapor deposition can be done, the ease with which electrical connections can be made, and lower energy and material requirements, amorphous modules can be mass produced much more quickly and cheaply than crystalline modules.[51]

Progress in a-Si technology has been rapid. While the first cells produced in 1976 were only 1 percent efficient, efficiencies have increased steadily, reaching in 1987 almost 14 percent for small area laboratory cells and almost 12 percent for larger area laboratory modules. (See Figure 5, left.) One of the leading scientists involved in a-Si development projects that an efficiency of 18 percent will be achieved in the laboratory in the early 1990s.[52]

Manufacturers in Japan, the United States, and Europe were quick to recognize amorphous silicon's potential as an inexpensive solar cell material. In a mere ten years, the a-Si solar cell has grown from a laboratory device to a commercial product,* for which production has steadily increased, reaching 11.9 MW or 41 percent of the total worldwide PV market in 1987.[53] (See Figure 5, right.)

---

* A key event in the early development of a-Si solar cells was the Japanese creation of the solar calculator market. Initially too expensive and inefficient to compete with crystalline silicon for the established PV power markets, amorphous silicon found a niche by replacing calculator and watch batteries. The use of a-Si solar cells has now expanded beyond these initial consumer electronics markets into such other applications as solar-powered battery chargers for vehicles, outdoor lights and electric fences, remote power systems, and utility power plants providing peak power during the daytime.

Box 1. Mining the Earth's Crust for Energy: Nuclear Power vs. a-Si Solar Power

The allure of nuclear power is the enormous amount of energy released in the fissioning of nuclear fuel. If breeder reactors were one day to become the dominant technology for nuclear power, repeated fuel recycling would make it feasible to fission about half the mass contained in the nuclear fuel. If the energy released in fission were converted to electricity at 33 percent efficiency, some 3800 kWh could be obtained from a single gram of uranium or thorium "feedstock"—nearly half the electricity consumed in a year by the average U.S. household. This potential is so large that it might lead to "mining the rocks" for uranium and thorium fuel. Even though the combined concentration of uranium and thorium in the crust averages just 50 parts per million, the amount of electricity that could be extracted from a tonne of this crust material would be equivalent to that which could be obtained by burning 70 tonnes of coal.

How does amorphous silicon solar cell technology compare? Even though the energy released in the fissioning of a single uranium nucleus is 100 million times greater than the "energy released" when a photon is absorbed in amorphous silicon, a uranium atom can fission only once, whereas a silicon solar cell can repeatedly absorb photons and convert solar energy into electricity. An amorphous silicon solar cell contains an amorphous silicon layer about 1 micron thick, amounting to some 3 grams of silicon per square meter of cell area. A 15-percent efficient PV system operated in the southwestern United States, where the insolation averages about 250 Watts per square meter, would thus produce about 3300 kWh per gram of silicon over the expected thirty-year PV system life—about the same as the amount of electricity from a gram of nuclear feedstock using breeder reactors.

But silicon is more than 5000 times more abundant than uranium and thorium in the earth's crust. In fact, it accounts for about half the mass of ordinary sand. (For perspective, the amount of electricity that could be produced from amorphous silicon cells using a tonne of sand is equivalent to what could be produced by burning more than a half a million tonnes of coal.)

While the first cells produced in 1976 were only 1 percent efficient, efficiencies have increased steadily, reaching in 1987 almost 14 percent for small area laboratory cells and almost 12 percent for larger area laboratory modules.

Present indications are that this rapid growth will continue. Chronar Corporation of Princeton, New Jersey, started building a 10-MW per year a-Si solar cell factory in late 1988[54] and has recently announced plans to build a 50-MW a-Si power plant in California.[55] Solarex, of Newtown, Pennsylvania, will have a 1-MW per year, fully computer-integrated manufacturing line operating in 1989 and is planning to build a 10-MW per year production facility in the early 1990s.[56] ARCO Solar, Inc., of Chatsworth, California, has designed a plant capable of 70 MW of yearly production.[57]

Today a-Si solar modules are manufactured in relatively small factories, with about 1 MW

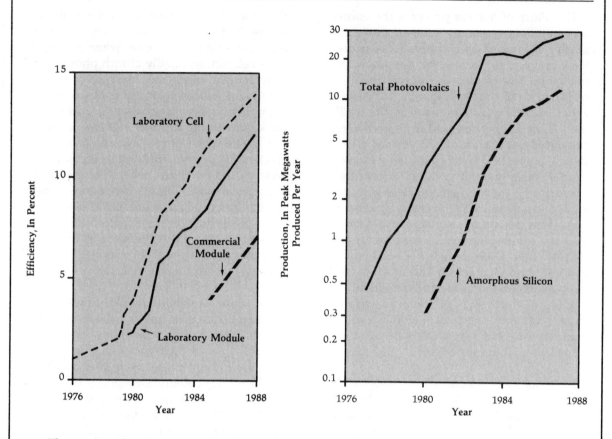

Figure 5: Progress in Amorphous Silicon Solar Cell Technology.

The trends in the efficiency of amorphous silicon (a-Si) small-area laboratory cells (typically 1cm x 1cm in size), larger area laboratory modules (about 1000 square centimeters), and commercially available modules are shown vs. calendar year in the left graph.

The right graph shows the global annual PV production volume and the global annual production of amorphous silicon solar modules in megawatts of peak power manufactured per year.

Sources: Y. Kuwano, SANYO, Private Communications, 1985; Z Erol Smith, Private Communications, 1987; D. Carlson, Solarex, Private Communications, 1987; *Photovoltaic Insider's Report,* February 1987 and January 1988.

per year production capacity at costs of about $1.5 to $1.6 per peak Watt for 5 to 7 percent efficient modules.[58] Amorphous silicon solar cells already cost less than other solar cell technologies—about half as much per peak Watt as polycrystalline cells and one third as much as single-crystal cells. Production costs of

a-Si solar cells are expected to decline further as the scale of plant production increases. *(See Figure 6.)*

Within a few years, Chronar's and Solarex's 10-MW per year plants are expected to produce 6-percent efficient cells for a cost of about $1

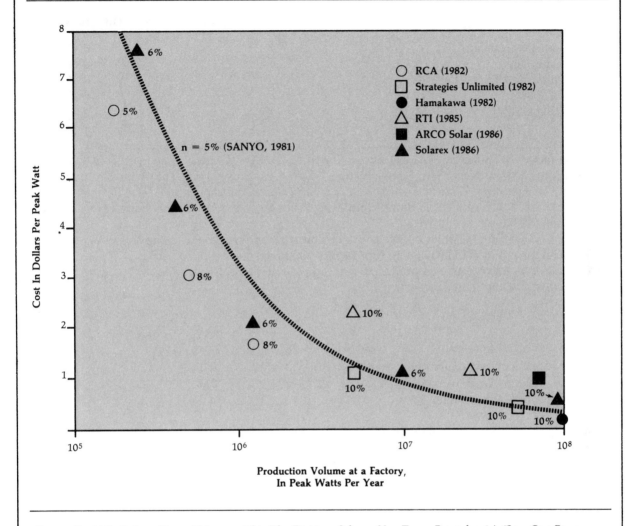

Figure 6: Estimated Manufacturing Cost for Amorphous Silicon Solar Modules at Various Factory Production Levels, Based on Recent Projections by Various Manufacturers and Researchers.

Source: David E. Carlson (General Manager, Thin Film Division, Solarex, New Town, Pennsylvania), "Low-Cost Power From Thin Film Photovoltaics," in *Electricity: Efficient End-Use and New Generation Technologies and Their Planning Implications,* T.B. Johansson, B. Bodlund and R.H. Williams, eds., Lund University Press, Lund, Sweden, 1989.

per peak Watt,[59] according to projections made by these companies. *(See, for example, Table 2.)* The cost of 5-percent efficient solar modules for the 50-MW Chronar power plant is expected to be $1.25 per peak Watt.[60] As amorphous silicon continues to move into the potentially vast electric power market, competition between

photovoltaic manufacturers in the United States, Japan and Europe is expected to set a fierce pace of innovation.

Although the eventual efficiencies and costs of amorphous silicon solar cells cannot be forecast with certainty, commercial module efficien-

Table 2. Estimated Amorphous Silicon Solar Module Production Cost for a Factory Producing 10 MW$_p$ per Year of 6-Percent Efficient Solar Modules[a]

	Dollars per W$_p$
Equipment Depreciation[b]	0.41
Direct Materials[c]	0.46
Direct Labor and Fringe Benefits[d]	0.16
Indirect Labor[e]	0.07
Indirect Expenses	0.06
Total	1.16

a. For a factory planned by Solarex. *Source:* David E. Carlson (General Manager, Thin Film Division, Solarex, Newtown, Pennsylvania), ''Low-Cost Power from Thin-Film Photovoltaics,'' in *Electricity: Efficient End-Use and New Generation Technologies and Their Planning Implications,* T.B. Johansson, B. Bodlund, and R.H. Williams, eds., Lund University Press, Lund, Sweden, 1989.

b. For a five-year depreciation period and a capital cost for equipment (computer-integrated manufacturing) in a 20,000 square foot facility estimated to be $16,500,000.

c. For 1 foot × 4 foot modules produced with an overall yield of 84%. The following is a breakdown of the materials cost:

Material	Cost ($/W$_p$)	Cost ($/m²)
Glass (chemically strengthened)	0.23	13.8
Silane	0.07	4.2
Encapsulant	0.04	2.4
Frame	0.03	1.8
Diborane, phosphine	0.03	1.8
Stannic chloride	0.02	1.2
Wire, other process gases	0.02	1.2
Aluminum	0.02	1.2
Total	0.46	27.6

d. 83 direct employees, 5-day work week, 2.5 shifts per day.

e. 17 indirect employees.

cies of 10 percent and 13 percent will almost certainly be achieved for single-layer and multi-layer cells, respectively, sometime between 1990 and 1995. (Multi-layer cells are made by depositing several thin-film layers, with each layer tuned to absorb a different part of the solar spectrum. Multi-layer cells are more efficient than single-layer cells, because they uti-lize more of the sun's spectrum.) In the longer term, it is likely that ''practical limit'' efficiencies of 12 to 14 percent will be achieved for single-layer amorphous silicon modules, while efficiencies of 18 to 20 percent are likely for multi-layer modules. By the turn of the century, PV module efficiencies of 12 to 18 percent (corresponding to PV system efficiencies of 10

Table 3. Cost and Efficiency Projections for Amorphous Silicon Solar Cells

	Single layer	Multi-layer
Efficiency:		
1987 small area cell (best laboratory results for 1 cm² device)	12%[a]	13%[b]
1987 submodule (laboratory module with area of about 1000 cm²)	8–9%[a]	12%[b]
1987 commercial module (typical, mass-produced)	5–7%[c]	
Projected 1990–1995 module	10%[d]	13%[d]
Practical limit	12–14%[e]	18–20%[e]
Costs (in $ per peak Watt):		
1988 Module manufacturing cost	1.5–1.6[f] [for 6–7% efficient modules[f]]	
Projected 1990s module manufacturing cost	0.6–1.2[g] [for 6–12% efficient modules[g]]	
Projected 2000 module manufacturing cost	0.2–0.4[g] [for 12–18% efficient modules]	

a. D.E. Carlson, Solarex, quoted in *Photovoltaic Insider's Report,* December 1987, p. 4.
b. D.L. Morel, ARCO Solar, quoted in *PVIR,* December 1987, p. 4.
c. *PVIR,* May 1987, p. 1.
d. Four U.S. manufacturers of a-Si solar cells (Chronar, Solarex, ARCO Solar and ECD) are in a cost-shared program with the U.S. Department of Energy to produce modules of these efficiencies by 1990, *PVIR,* February 1987, p.2.
e. D.E. Carlson, Solarex, private communication, 1987.
f. For a plant producing 1.3–1.4 MW_p per year. S. Kaplan, Chronar Corp., private communications, 1988. 1988 prices are about $4–$5 per peak watt.
g. Based on estimates by amorphous silicon solar cell manufacturers for factories producing 10–300 MW_p of solar cells per year.

to 15 percent) and costs of $0.20 to $0.40 per Wp may well be achieved. *(See Table 3 and Box 2.)*

To indicate briefly how these manufacturing costs might be reduced, consider possible improvements on the Solarex design for a

Box 2. Key Technical Assumptions Relating to Amorphous Silicon Solar Cells

This analysis of the prospects for PV hydrogen is based on the assumption that the cost and performance goals for amorphous silicon solar cells near the turn of the century will be realized. Here is the technical basis for these goals.

Assumption #1: *Amorphous silicon PV modules will have stable efficiencies of 12 to 18 percent.*

While today's commercially available single-layer amorphous silicon modules have modest conversion efficiencies of about 5 to 7 percent, efficiencies achieved in the laboratory with small-area, multi-layer cells have exceeded 13 percent. To date, efficiencies achieved in the laboratory have been realized in commercial modules some five to six years later (*Figure 5, left*), but this time lag is expected to shorten in the future, as manufacturers move toward computer-integrated manufacturing of a-Si cells and achieve generally higher levels of quality control in production than in the laboratory—a phenomenon that has occurred for a variety of products in the semiconductor industry.[67] By the early to mid–1990s, efficiencies for commercial modules are expected to be 10 percent for single-layer cells and 13 percent for multi-layer cells.[68] One of the leading scientists in amorphous silicon solar cell research projects that efficiencies of 18 percent will be achieved in the laboratory by the early 1990s.[69] By the year 2000, commercial modules approaching the ''practical limit'' values of 12 to 14 percent for single-layer solar cells and 18 to 20 percent for multi-layer cells may well become available.[70,71]

One problem that plagued the early development of amorphous silicon solar cells is that the cells experience an initial loss of efficiency (known as the Staebler-Wronski effect[72]) when exposed to light. While the Staebler-Wronski effect is not yet fully understood theoretically, the problem has been largely solved in practice. The initial efficiency degradation can be completely reversed by reheating the cells to their annealing temperature (about 200 °C) for a few minutes. At typical outdoor solar cell operating temperatures of 50 °C to 60 °C partial annealing takes place, which tends to counterbalance the Staebler-Wronski effect.[73] Also, making single-layer cells thinner and using multiple layers tends to retard the initial loss of efficiency. Single-layer and multi-layer modules can now be made that stabilize, after a few months exposure to sunlight, at about 80 percent and 90 percent of their initial efficiencies, respectively.[74]

Given these results, the assumption in this analysis is that stabilized solar module efficiencies of 12 to 18 percent will be reached around the turn of the century. We further assume that the efficiency of a large PV system without power conditioning is 85 percent of the individual module efficiency, due to electrical losses in wiring and perhaps to wind-blown dirt or dust on the modules.[75] Thus, a PV system constructed of 12- to 18-percent efficient modules would have an overall efficiency of 10.2 to 15.3 percent.

Assumption #2: *The production cost of amorphous silicon PV modules will be in the range $0.2 to $0.4 per peak Watt.*

Today most commercial amorphous silicon solar cells are produced in batch operations in small facilities with production capacities of the order of 1 Megawatt peak per year or less. The current production cost is estimated at $1.5 to $1.6 per peak Watt for 6-percent efficient cells.[76] In larger plants, considerable economies of scale could be realized. According to Dr. E.S. Sabisky,

Box 2. (cont.)

former manager of the Amorphous Silicon Research Project, at the Solar Energy Research Institute in Golden, Colorado, now at Chronar:[77]

"...if today's thin-film amorphous silicon modules of 6 to 8 percent efficiency are combined with a 10 megawatt annual production plant, the module cost target of $1 per peak Watt can be reached..."

Sabisky's prognosis is reflected in the announcement in September 1988 that Chronar, of Trenton, New Jersey, will build a 50-MW plant site 60 miles north of Los Angeles for $125 million.[78] This plant, which will sell the produced electricity to the Southern California Edison Company, represents an enormous scale-up from the largest amorphous silicon facility built to date, a 100 kW generating field operated by Alabama Power Company. The PV modules in this plant are expected to have initial efficiencies of 7.5 percent and guaranteed stabilized efficiencies of 5 percent and are expected to cost $1.25 per peak Watt.[79] Similar cost estimates have been projected by other manufacturers and are shown in Figure 6.

In Table B.2.1, we show how costs might evolve from current levels to the range of $0.2 to $0.4 per Wp. These estimates are based on a detailed cost evaluation by Solarex researchers for a 10-MW per year plant, which would produce 6-percent efficient modules costing $1.16 per peak Watt.[80] *(See Table 2.)* Table B.2.2 shows the corresponding costs for AC and DC electricity and hydrogen derived from this electricity through electrolysis.

Increasing the scale of production from 10 MW to 100 MW per year would lead to savings mainly in labor costs, and the cost of 6-percent efficient modules would be reduced from $1.16 to $0.94 per peak Watt.

Increasing the efficiency of thin-film solar cells generally involves fine tuning the composition of the active materials—altering the thickness of the active layer, adjusting the amounts of dopants, adding additional layers that use different parts of the solar spectrum, adjusting the alloy composition of the cells used in different layers, etc. These changes probably involve little or no increase in the labor and materials costs per unit area.[81] Thus, a doubling or a tripling of cell efficiency would probably lead to a twofold or three-fold reduction in the cost per peak Watt.

Materials costs could also be reduced. For the Solarex design, glass accounts for $0.23 per peak Watt or half of the materials cost. The cost of glass could be reduced to about $0.06 per peak Watt, if chemical strengthening were not required. A further reduction in the cost of glass could be realized in large-scale (>60 MW) facilities, where it would be possible to integrate a float-glass manufacturing plant with an a-Si plant. Recovering silane (SiH_4) gas (the primary feedstock for amorphous silicon deposition) during processing, and reducing module framing costs could further reduce material costs. Researchers at Solarex estimate that the overall cost of materials could potentially be reduced to about $0.11 per peak Watt through innovations such as these, for 12-percent efficient cells produced in a 200-MW per year production facility.

Finally, as the technology matures, the rate of equipment obsolescence will slow, making it possible to increase the equipment depreciation period. The effect of increasing the depreciation period from five to ten

Box 2. (cont.)

years plus the effects of the materials innovations mentioned above are summarized for a 1.67 million square meters per year production facility in the last three columns of Table B.2.1. Note that total production costs with such innovations would be in the range $0.16 to $0.25 per peak Watt for 12- to 18-percent efficient modules, somewhat lower than the range we assume in this analysis.

Assumption #3: Amorphous silicon PV modules will have lifetimes of 30 years.

Because amorphous silicon solar cells are a new technology, field tests of more than a few years have not yet been completed. However, present-day commercial modules pass a battery of accelerated environmental tests. These tests are designed to simulate many years of use in a short time by subjecting the solar modules to rapidly varying extremes of light, temperature, humidity, hail impacts, etc. A preliminary judgment (which must be verified by further field testing), based on the results of such tests and

expected processing improvements, is that a 30-year lifetime is a reasonable expectation.[82]

Assumption #4: Area-related balance of system costs will be $33 per square meter for large fixed flat-plate amorphous-silicon-based PV arrays.

Area-related balance of system (BOS) costs include the support structure holding the PV modules, the array wiring and electrical equipment, land, site preparation and other construction costs. Previous conceptual design studies and analysis of data from experimental PV arrays and demonstration projects indicate that area-related BOS costs of $50 per square meter could be readily achieved with present technology.[83,84] If low-cost support structures using pre-fabricated PV panels were employed, this cost could be reduced to perhaps $37 per square meter. With a low-current, high-voltage electrical design, which is especially well suited to amorphous silicon cells, wiring costs could be reduced to give a total area-related BOS cost of $33 per square meter, as assumed in this study.[85]

10-MWp per year production facility that could be built with present technology, producing 6-percent efficient cells costing $1.16 per Wp. *(See Table 2 and Box 2, Tables B.2.1 and B.2.2.)* If the output of the facility were expanded to 100 MWp per year the production cost could be reduced to $0.94 per Wp, largely as the result of savings in labor at the larger scale. For amorphous silicon solar cells, the cost per square meter is not expected to change much with the efficiency because the active materials in the cell account for only a few percent of the total cost *(See Table 2)* and because efficiency improvements will be achieved by fine-tuning the composition of the active materials—

altering the thickness of the active layer, adjusting the amounts of dopants, adding additional layers that use different parts of the solar spectrum, adjusting the alloy composition of the different layers, etc. Accordingly, the cost for 12-percent efficient cells, which are expected to be commercially available by 1995, should be about $0.47 per Wp. The technology would be sufficiently mature with 12-percent efficient modules that the depreciation period for the PV production facility could probably be extended from five years (assumed for the 10-MW per year case producing 6-percent efficient cells shown in Table 2) to ten years—thereby reducing the cost of cells further, to

Table B.2.1. Production Cost of Amorphous Silicon Solar Cells ($ per W$_p$)

Annual Production Million sq. m.	5-Year Depreciation						10-Year Depreciation w/Reductions in Materials' Costs[c]		
	0.167			1.67			1.67		
MW$_p$	10	20	30	100	200	300	100	200	300
Efficiency	6%[a]	12%[a]	18%[b]	6%[a]	12%[a]	18%[b]	6%	12%	18%
Variable Costs									
Direct Labor	0.16	0.08	0.05	0.06	0.03	0.02	0.06	0.03	0.02
Materials	0.46	0.23	0.15	0.46	0.23	0.15	0.22	0.11	0.07
Fixed Costs									
Indirect Labor	0.07	0.035	0.023	0.01	0.005	0.003	0.01	0.005	0.003
Indirect Expenses	0.06	0.03	0.02	0.01	0.005	0.003	0.01	0.005	0.003
Depreciation	0.41	0.205	0.14	0.40	0.20	0.133	0.20	0.10	0.065
Total Cost	1.16	0.58	0.39	0.94	0.47	0.31	0.50	0.25	0.16

a. These estimates were made by Carlson [David E. Carlson (General Manager, Thin Film Division, Solarex, Newtown, Pennsylvania), "Low-Cost Power from Thin-Film Photovoltaics," in *Electricity: Efficient End-Use and New Generation Technologies and Their Planning Implications*, T.B. Johansson, B. Bodlund, and R.H. Williams, eds., Lund University Press, Lund, Sweden, 1989], based on computer integrated manufacturing technology under development at Solarex, in Newtown, Pennsylvania. In this analysis, detailed cost estimates were made for present-day production technology using 6% efficient cells, and these cost estimates were extrapolated to 12% efficient cells, assuming that the costs per unit area will not change as the efficiency is increased.

b. These cost estimates are obtained by extrapolating Carlson's estimates to 18% efficient cells, with the assumption that the cost per unit area will not change as the efficiency is increased.

c. Two possibilities for cost reduction are taken into account here: (i) reduced materials costs and (ii) an extended equipment depreciation period. Carlson estimates that the collective effect of the materials cost reduction efforts discussed in the text would be to reduce the cost of materials from $0.46 per peak Watt for 6% efficient cells manufactured at a rate of 10 MW per year to perhaps $0.11 per peak Watt for 12% efficient cells produced at a rate of 200 MW per year (D.E. Carlson, "Low-Cost Power from Thin-Film Photovoltaics"). Since the pace of innovation will probably slow as the technology matures, it is assumed here that the depreciation period is doubled, from five to ten years.

$0.37 per Wp. The prospects are also good for reducing the materials costs associated with cell manufacture. First, with a larger production facility it becomes feasible to couple a PV production facility to a dedicated glass production plant. Plant integration together with reductions in silane gas losses in a-Si cell processing, elimination of the need for chemically strengthening the glass, and reductions in module framing costs could collectively cut

Table B.2.2. Production costs for PV electricity and H2 as technology advances

	PV Modules[a] ($ per W_p$)	Electricity[b] (cents per kWh)		Hydrogen[c] ($ per GJ)
		AC	DC	
1. Present technology: 6% modules, $50/m² BOS, 10 MW_p/yr output	1.16	11.09	10.05	35.8
2. Increase output, 10 to 100 MW_p/yr	0.94	9.98	8.98	32.2
3. Increase module efficiency, 6 to 12%	0.47	5.30	4.49	17.7
4. Increase depreciation period for factory equipment, 5 to 10 years	0.37	4.79	4.01	15.8
5. Reduce materials cost for module, $27.6 to $13.2/m²	0.25	4.18	3.43	13.9
6. Reduce BOS cost, $50 to $33/m²	0.25	3.47	2.74	11.6
7. Increase module efficiency, 12 to 18%	0.16	2.49	1.79	8.4

a. *See Table B.2.1.*
b. The PV electricity cost (in $/kWh) is (neglecting taxes, subsidies):
   $C = (CRF + INS) \times (1+ID) \times [(Cmod \times Ip + Cbos/nmod)/nsys + Cpc \times Ip]/(insol \times 8760) + Com/(nmod \times nsys \times insol \times 8760)$
   CRF = capital recovery factor = $i/[1 - (1 + i)^{-N}]$
   i = real discount rate = 0.061 (appropriate for utility investments in U.S.)
   N = equipment lifetime = 30 years
   INS = insurance rate = 0.005 (from EPRI *Technical Assessment Guide*)
   ID = indirect cost factor = 0.25 (based on Sandia experience)
   Cmod = capital cost of modules ($/kW)
   Cbos = area-related balance-of-system capital cost ($/m²)
   Ip = maximum insolation = 1 kW/m²
   nmod = solar module efficiency
   nsys = system efficiency (excl. module) = 0.85 (DC); = 0.96 × 0.85 =0.816 (AC)
   Cpc = capital cost for power conditioning = $150/kW (AC); = 0 (DC)
   insol = annual average insolation on tilted array =0.271 kW/m² (El Paso, TX)
   Com = annual O&M cost = $0.45/m²/year (based on ARCO Solar experience)
c. For an 84% efficient electrolyzer, the cost of hydrogen (in $/GJ) is Ch = 2.52 + 3.307 × Cdc, where Cdc is the DC electricity cost, in cents/kWh.

materials costs in half and thus reduce the total cost of modules to $0.25 per Wp.* Even further cost reductions could be achieved with commercial module efficiencies higher than 12 percent; for instance, with 18-percent efficient modules, the cost would be $0.16 per peak Watt. *(See Table B.2.1.)*

*Some of the most significant opportunities indicated here for reducing the cost of materials have already been adopted at Solarex since this analysis was carried out—e.g. the elimination of chemical strengthening of the glass and the introduction of a silane recovery process.

Implications for the Future of Photovoltaic Power

What are the implications of these advances for stationary power generation? Table 4 shows markets for PV electricity, with the "break-even" solar module price needed to compete with existing sources of electricity. In recent years, the drop in prices (to current levels of about $4 to $5 per Wp) has permitted PV power to move beyond small, highly specialized markets into water pumping and small-scale power-generating applications. In sunny areas outside the reach of centralized power sources, such as some remote military bases, research stations and rural areas in developing countries,[61] PV power has become economically competitive with the small Diesel generators now used to generate electric power.

By the 1990s, when production costs of amorphous silicon solar modules are expected to be in the range $0.6 to $1.2 per Wp, amorphous silicon PV arrays could start replacing oil and gas-fired generators for electric utility peaking applications. (Chronar's planned 50-MW plant will essentially be a peaking plant, the electricity from which will be sold to the Southern California Edison Company.) In the United States, utility peaking will very likely be the most important near-term PV market,[62] followed by daytime power generation on residential rooftops. *(See Table 4.)*

Although these significant local markets will be important for the PV industry, their total impact on global energy supply will be small.[63] For PV power to make major inroads into the electric power market, PV electricity would have to be made available on demand, which means that PV power must not only become very cheap but also that inexpensive methods must be found for storing electricity for use when the sun doesn't shine.[64,65]

As shown in Chapter 4, it is likely that DC electricity produced in the southwestern United States could cost as little as $0.02 to $0.035 per

kWhDC by the year 2000, if goals for amorphous silicon solar cells are achieved. If PV DC electricity could be produced for say $0.02 per kWhDC, then PV baseload power could cost about $0.05 per kWhAC with advanced batteries or underground pumped hydro storage systems, making it competitive with baseload power from coal-fired and nuclear power plants.

To reach such low electricity costs with PV would, however, require not only achieving technological goals (18-percent efficient PV modules costing $0.20 per Wp and balance of systems costs of $33 per m²), but also the right location, e.g. one with a good solar resource. (A PV electric system must be located relatively near the demand, because it is very costly to transmit power over long distances with present technology.) In much of the United States, the insolation is simply too low to reach these costs.

Even in sunny areas where storage technologies could be developed, the overall potential impact of PV electricity would be limited by the restricted role of electricity in the overall energy economy. Electricity consumption accounts for only about ten percent of global energy use, and electricity production for only about one fourth of all fossil fuel use.

Clearly, PV technology could have a much greater impact in reducing fossil fuel use if economical ways could be found for making synthetic fluid fuels with solar cells as direct substitutes for oil and gas, which account for two thirds of global fossil fuel use.

Clearly, PV technology could have a much greater impact in reducing fossil fuel use if economical ways could be found for making syn-

Table 4. Potential Markets for Photovoltaic Electricity

Market	Breakeven Solar Module Price ($ per peak Watt)[c]	Potential Market Size Total (MW$_p$)	Annual (MW$_p$/year)
Corrosion protection[a]	20–100[d]		
Buoys[a]	60[d]		
Consumer products: calculators, watches, etc.[a]	10[d]		9[e]
Remote water pumping[a]	4–7[d]	2,000[f]	100[b]
Diesel generator replacement, remote power[a]	5[d]	10,000[g]	500[b]
U.S. Utility electric peaking	2–3[d]	50,000 (Total U.S. Peaking)	2,500[b]
Daytime power for grid-connected residences in U.S.[h]	0.7–1.5	100,000	5,000[b]
U.S. Utility electric baseload (with storage)[i]	0.15–0.45	600,000 (Total U.S. Baseload)	30,000[b]

a. Currently commercial markets. The total PV market in 1987 was about 29 MW.
b. The annual potential market was estimated by dividing the total potential market by 20 years commercialization time.
c. Breakeven PV prices were estimated assuming 10% efficient PV systems, and balance of systems costs of $33 to $100 per square meter, depending on system size.
d. P.D. Maycock and E.N. Stirewalt, *The Photovoltaic Revolution*, Rodale Press, NY, NY, 1985.
e. Estimated size of 1986 consumer products market for solar cells, *Photovoltaic Insider's Report*, July 1987, p. 2.
f. Assuming that 2 million 1 kW water pumping systems are installed.
g. Assuming that 2 million remote sites acquire 5 kW PV systems for communications, refrigeration and lighting.
h. Assuming that the PV systems must compete with residential grid power costing $0.06/kWh, and that 50 million residences acquire 2 kW PV systems.
i. Assuming that baseload PV electricity with storage costs $0.08/kWhAC.

thetic fluid (liquid or gaseous) fuels with solar cells as direct substitutes for oil and gas, which account for two thirds of global fossil fuel use. As many researchers have pointed out, it is technically feasible to use hydrogen produced by splitting water through PV-powered electrolysis to replace oil and natural gas in virtually all their present uses.[66] Making hydrogen this way provides a means of storing solar energy, and hydrogen produced in sunny areas can be transported long distances to remote markets at much lower costs than electricity. And PV hydrogen has obvious environmental advantages.

IV. Designing a PV Hydrogen Energy System

While PV is environmentally preferable to alternative fuels, to be economically acceptable it must also be able to provide the consumer with energy services (transport, heating, etc.) at costs comparable to those associated with alternative synthetic fuels. Here the design of a PV hydrogen system is described, based on a-Si solar cells with the aim of better understanding the prospects for making PV hydrogen economically competitive.

A PV hydrogen production system would consist of four major parts. It would have an array of solar modules that would collect sunlight and convert it into DC electricity, an electrolyzer that would use this electricity to split water into hydrogen and oxygen, a compressor that would bring the produced hydrogen up to the required pressure, and a hydrogen storage unit. If the hydrogen production exceeds onsite demand, the system would also include a pipeline that would carry the excess hydrogen to remote users. *(See Figure 4.)*

Finding Sites for PV Hydrogen Production

Where should PV hydrogen systems be located? Would it be better to produce the electricity locally, thereby avoiding the expense of long-distance transport? Or should the electricity be produced primarily in the sunniest areas, making it possible to extract more useful energy from a given solar cell and thereby reducing the cost of PV electricity?

The choice is not a difficult one. Even for the PV DC electricity production costs in the range $0.020 to $0.035 per kWh expected for sunny areas near the turn of the century, electricity production for electrolysis would still account for 60 to 70 percent of the cost of hydrogen delivered to consumers a thousand miles from the production site. *(See Figure 7.)* Economics thus dictate that a PV hydrogen economy would involve locating solar hydrogen production systems in sunny areas like the southwestern United States, where the average insolation is typically at least 50 percent higher than in much of the rest of the country. *(See Table 5.)* The hydrogen could then be piped to areas with less sunshine.

Transporting hydrogen long distances via pipelines typically costs far less than transporting the same amount of energy in the form of electricity via transmission lines.[86] While it would not make sense to generate PV electricity in sunny regions for transmission to distant demand areas, at least with present technology, it would make sense to develop PV hydrogen systems this way, because long-distance hydrogen transmission is relatively inexpensive. (As Figure 7 shows, hydrogen transmission costs are a relatively small part of the total cost of hydrogen.)

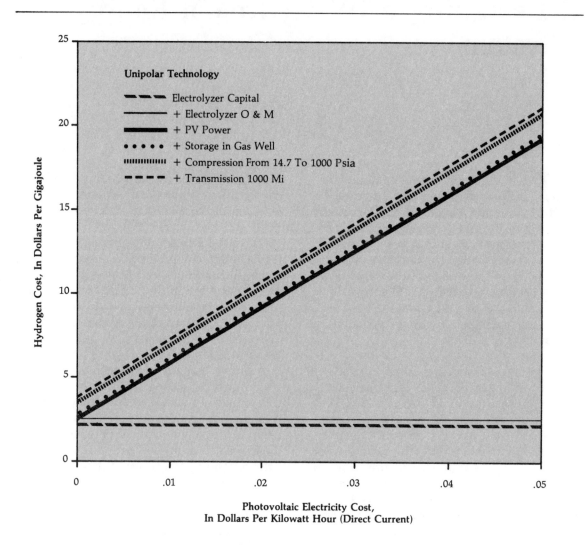

Figure 7: The Production Cost of Photovoltaic Hydrogen versus Direct Current Electricity Cost for Unipolar Technology.

Unipolar Technology

- – – – Electrolyzer Capital
- ——— + Electrolyzer O & M
- ——— + PV Power
- • • • • • + Storage in Gas Well
- ||||||||||| + Compression From 14.7 To 1000 Psia
- – – – – + Transmission 1000 Mi

Hydrogen Cost, In Dollars Per Gigajoule

Photovoltaic Electricity Cost,
In Dollars Per Kilowatt Hour (Direct Current)

The hydrogen production cost includes the electrolyzer capital cost, the electrolyzer operation and maintenance costs, the cost of PV DC electricity, the cost of storage, the cost of compression from the electrolyzer pressure of 14.7 psia to pipeline pressure of 1000 psia, and the cost for 1000 miles of pipeline transmission.

Designing the PV Electricity Production System

Once a site has been selected, the next task is to design a PV array to produce low-cost electricity for electrolytic hydrogen production. Since solar cells produce the direct current (DC) electricity required by electrolyzers for PV hydrogen production, the extra cost of DC to AC conversion equipment is avoided. Moreover, there is good theoretical and empirical

Table 5. The Solar Resource in Various U.S. Cities[a]

City	Latitude (degrees)	Annual Average Insolation (Watts/m²) Incident on a Flat Plate PV Array	
		Horizontal Surface	Surface Tilted at Latitude Angle
Las Vegas, Nevada	36.1 N	245	273
El Paso, Texas	31.8 N	250	271
Phoenix, Arizona	33.4 N	246	267
Tucson, Arizona	32.2 N	246	267
Albuquerque, New Mexico	35.1 N	236	266
China Lake, California	35.7 N	241	263
Bakersfield, California	35.4 N	230	245
Amarillo, Texas	35.2 N	218	241
Pueblo, Colorado	38.3 N	213	241
San Diego, California	32.7 N	210	228
Los Angeles, California	33.9 N	209	227
New Orleans, Louisiana	30.0 N	189	198
Atlanta, Georgia	33.7 N	177	187
Chicago, Illinois	41.8 N	160	172
Newark, New Jersey	40.7 N	153	166
Boston, Massachusetts	42.4 N	145	157
Portland, Oregon	45.6 N	140	147

a. *Source:* Jan F. Kreider and Frank Kreith, *Solar Heating and Cooling: Active and Passive Design,* McGraw-Hill Book Company, New York, NY, 1982.

evidence from both the United States and Germany that a direct connection between the solar array and the electrolyzer works extremely well and obviates the need for expensive DC to DC power-conditioning equipment.[87]

While most research in PV power development to date has been on reducing the cost of the PV modules themselves, some efforts have been aimed at developing low-cost designs for the "balance of system" (BOS). The balance of system includes all the components in the PV system except the modules themselves: for example, the support structure holding the modules, the electrical wiring and equipment, land and construction. In general, the BOS cost has two components—an "area-related" cost and a "power-related" cost. In PV hydrogen production, all the BOS costs are area-related, since no power conditioning equipment is needed.

How large are BOS costs? Area-related BOS costs from several conceptual design studies (RCA, Sandia/Battelle, EPRI, USDOE) for flat-plate, tilted fixed arrays[88] and field data from ARCO Solar and European Economic Community PV projects are summarized in Table 6. These costs vary from about $47 to some $230 per square meter of collector area.

The more efficient the solar modules, the smaller the area-related BOS costs, since less

Table 6. Area-Related Balance-of-Systems Cost Estimates for PV Arrays

		Conceptual Designs		Field Studies	
Component	RCA[a]	Sandia[b]	EPRI[c]	ARCO[d]	EEC[e]
Array support structure	41.2	26.2	46.2	48	30–100
Wire	3.2	12.7	5.65	65	
				(includes all electrical)	
				$15 subtracted for PCU	
Other DC electric	0.5	14.4	9.2		
Site-prep and cleanup	1.8	2.5	2.8	120	
				(includes engineering)	
Land[f]	0.2	0.25	0.25	0.25	–
Total BOS cost	47	56	64	233	50–180

a. J. Stranix and A.H. Firester, "Conceptual Design of a 50MW Central Station Power Plant," RCA Laboratories, Princeton, NJ, 1982.

b. G.T. Noel, D.C. Carmichael, R.W. Smith, and J.H. Broehl, "Optimization and Modularity Study for Large-Size PV Flat-Panel Array Fields," Battelle-Columbus, 18th IEEE PV Spec. Conf., Las Vegas, Nevada, October 1985.

c. S.L. Levy and L.E. Stoddard, "Integrated Photovoltaic Central Station Conceptual Designs," EPRI Report AP-3264, June 1984.

d. G.J. Shushnar, J.H. Caldwell, R.F. Reinoehl, and J.H. Wilson, "ARCO Solar Field Data for Flat-Plate PV Arrays," 18th IEEE PV Specialists Conference, Las Vegas, October, 1985.

e. G. Grassi, Commission of the European Communities, DC XII, Brussels, Belgium, 18th Photovoltaic Specialists Conference, Las Vegas, Nevada, October 1985; G. Grassi, P. Paoli, L. Leonardini, E. Vitali, P. Conti, E. Colpizzi, "Low-Cost Support Structure for Large Photovoltaic Generators," 18th IEEE Photovoltaics Specialists Conference, October 1985.

f. For land costing $1000/acre.

land, smaller support structures, and less wire are needed to produce the same amount of PV power. A BOS cost of $100 per square meter would contribute to the cost of electricity only 25 percent as much as a 10-percent efficient module at the 1987 module price of $5 per Wp. Consequently, the highest priorities in PV development today should still be reducing the module cost and increasing the module efficiency. But by the turn of the century, when the cost of 12- to 18-percent efficient modules is expected to be less than $0.50 per Wp, BOS cost reduction will become relatively more important. For example, if the cost of 12-percent efficient modules were $0.40 per Wp, the BOS contribution to the cost of electricity would be as large as the module contribution at $50 per square meter. Clearly, considerable emphasis should be given to finding low-cost BOS designs over the next decade or so.

Fortunately, much can be accomplished simply with "off-the-shelf" components and established construction practices. For this analysis a design was selected that brings together the best features of the various designs cited in Table 6. The estimated BOS cost of this design is $33 per square meter. Most of the savings come from adopting the low-cost Sandia/Battelle design for the collector array structure (which makes use of pre-fabricated panels) and the RCA electrical design (in which the system operates at high voltage and low current, thereby saving on the cost of the more expensive wire needed to carry a higher current, and coincidentally requiring fewer electrical connections).[89] (With innovative designs it may be possible to reduce BOS costs further—to $20 per square meter or less.[90])

In Figure 8, the cost of DC electricity from an a-Si solar array is plotted as a function of solar module and BOS costs for 12-percent efficient modules.[91,92] This graph can be used to esti-

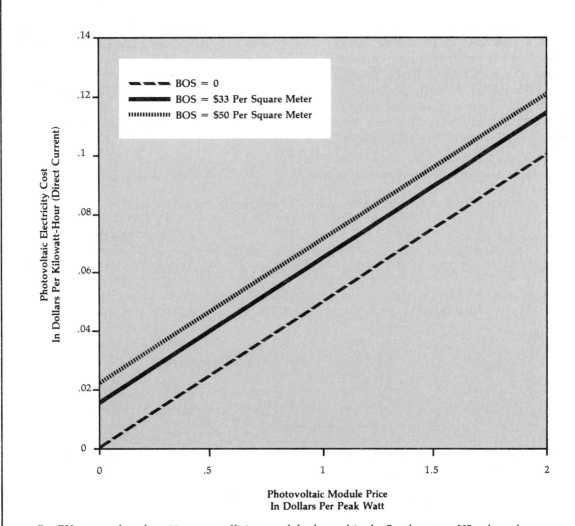

Figure 8: The Cost of PV DC Electricity as a Function of the Solar Module and Balance of Systems Costs.

- ▬ ▬ ▬ BOS = 0
- ▬▬▬ BOS = $33 Per Square Meter
- ▥▥▥▥ BOS = $50 Per Square Meter

Y-axis: Photovoltaic Electricity Cost In Dollars Per Kilowatt-Hour (Direct Current)

X-axis: Photovoltaic Module Price In Dollars Per Peak Watt

For PV systems based on 12 percent efficient modules located in the Southwestern US, where the average insolation is 271 Watts per square meter, assuming a system lifetime of 30 years and an annual operation and maintenance cost of $0.45 per square meter. (See Note 92.)

mate the cost of DC electricity in an area with insolation conditions like those in El Paso, Texas for modules with this efficiency. For example, with a BOS cost of $33 per square meter and a PV module cost of $0.40 per Wp, the cost of DC electricity would be about $0.035 per kWh. Alternatively, if a certain DC electricity cost is desired, a range of satisfactory PV module and BOS costs can be inferred. For example, to achieve a DC electricity cost of $0.025 per kWh with a BOS cost of $33 per m², the module cost would have to be $0.20 per Wp.

Table 7 shows the projected cost of DC electricity in the southwestern United States over time, based on the projections for a-Si solar technology shown in Table 3. These projections imply that the cost of DC solar electricity should fall from its present value of about $0.28 per kWh (based on the 1987 solar module selling price of $5 per Wp and an efficiency of

5 percent) to about $0.044 to $0.089 per kWh by 1990 to 1995 (for solar module costs of $0.6 to $1.2 per Wp and module efficiencies of 6 to 12 percent), and to the range $0.020 to $0.035 per kWh by 2000, if module costs fall to the range $0.20 to $0.40 per Wp and module efficiencies of 12 to 18 percent are achieved.

The Electrolytic Production of Hydrogen

The cost of electrolytic hydrogen depends on the capital cost of the electrolyzer plus the cost of the DC electricity needed to run it.[93,94] The cost of hydrogen from a large (>10 MW) electrolyzer is shown in Figure 7, as a function of the cost of DC electricity. Using PV electricity costs based on projections for a-Si solar cell technology, we can estimate the projected production cost of PV hydrogen by year. (See Table 7.)

Table 7. Projected Cost of PV Electricity and PV Hydrogen in the Southwest[a]

| Year | PV Module Cost ($/W_p$) | EFFICIENCY | | COST OF PRODUCTION | | |
		Module	PV system	PV Electricity ($/kWhDC)	PV Hydrogen ($/GJ)	($/gal)[b]
1988	5.0 (price)	5%	4.3%	0.279	95	11.6
	1.6 (cost)	5%	4.3%	0.114	40.2	4.9
1990–1995	1.2	6%	5.1%	0.089	31.8	3.88
	0.6	12%	10.2%	0.044	17.2	2.10
2000	0.4	12%	10.2%	0.035	14.0	1.70
	0.2	18%	15.3%	0.020	9.10	1.11

a. PV electricity and hydrogen production costs are calculated for a PV hydrogen system greater than 5 MW in size, assuming that the BOS costs are $33/m², the system lifetime is 30 years, the average annual insolation is 271 Watts/m², and the annual operation and maintenance costs are $0.45/m². It is assumed that a unipolar electrolyzer is used to produce hydrogen at atmospheric pressure. It is also assumed that the PV array is directly coupled to the electrolyzer at 93 percent coupling efficiency. The cost of hydrogen compression, storage, transmission, and local distribution are not included. (See Notes 92, 94.)
b. The cost of energy expressed in dollars per gallon of gasoline-equivalent.

The cost of hydrogen production for a range of smaller-sized systems is shown in Figure 9. While the unit cost of electrolytic hydrogen declines as plant size increases, there is little economy of scale for hydrogen production rates above about 2 MW.[95] Since there are likewise no significant scale economies for PV power production above production levels of some 5 to 10 MW,[96] hydrogen production operations could be highly modularized, with typical module capacity in the range 5 to 10 MW and characteristic capital costs of $4 million to $12 million for electrolyzer plus PV equipment (for 12- to 18-percent efficient PV modules costing $0.2 to $0.4 per Wp).

If hydrogen is to be produced in the Southwest and transmitted by pipeline to distant

Figure 9: The Cost of Electrolytic Hydrogen as a Function of Plant Size and DC Electricity Cost, for Unipolar Electrolysis at Atmospheric Pressure, with No Compression or Storage.

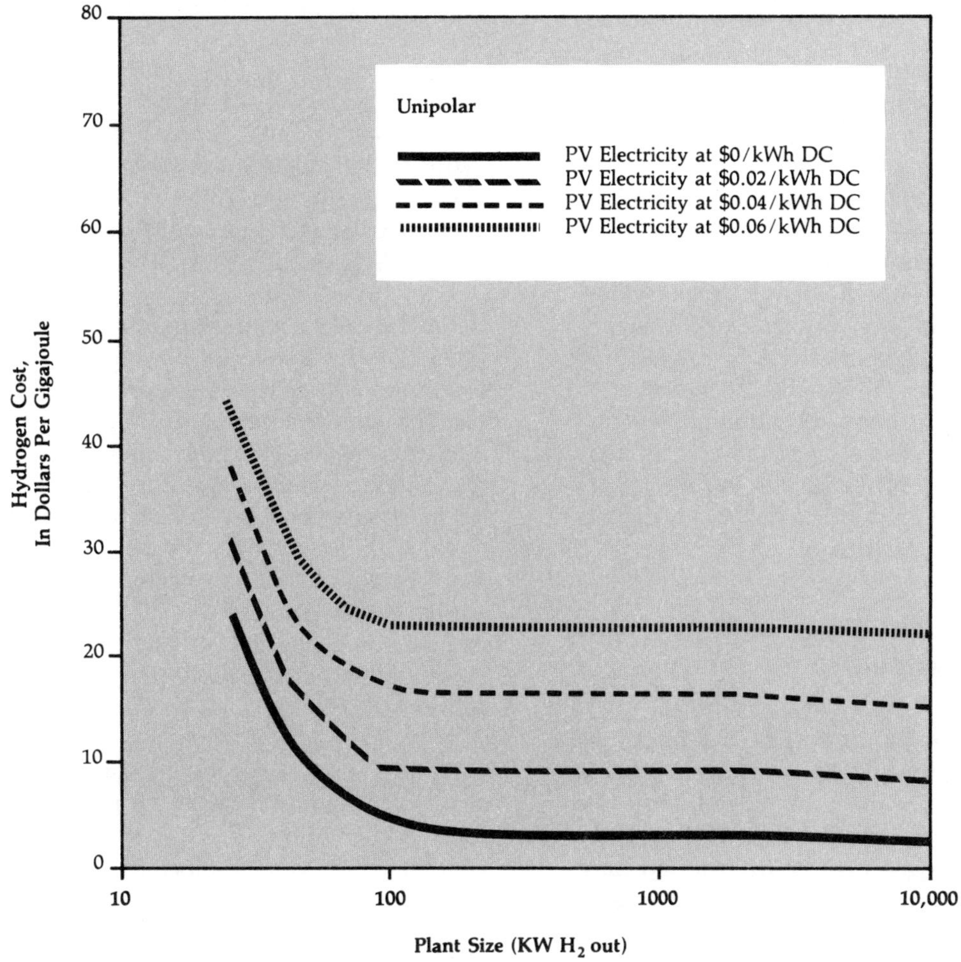

Unipolar

——	PV Electricity at $0/kWh DC
– – –	PV Electricity at $0.02/kWh DC
− − −	PV Electricity at $0.04/kWh DC
∣∣∣∣∣∣	PV Electricity at $0.06/kWh DC

Hydrogen Cost, In Dollars Per Gigajoule

Plant Size (KW H$_2$ out)

markets, two alternative electrolytic strategies could be pursued. One approach would involve making hydrogen at atmospheric pressure and then compressing it up to the high pressure needed to move the gas through the pipeline. An alternative approach would be to carry out the electrolysis at high pressure, thereby dramatically reducing the otherwise considerable amount of compressor work required. But while using pressurized electrolyzers lead to savings in compressor work, these devices tend to be slightly more costly and less efficient than their atmospheric pressure counterparts. As a result, the cost of hydrogen delivered at the end of long pipelines tends to be less with the more efficient, less capital-intensive atmospheric pressure units.

When water is split electrolytically oxygen, a valuable byproduct, is produced as well as hydrogen. If this oxygen could be sold in local markets the byproduct credit would be some $1.5 to $2.2 per GJ of produced hydrogen[97]— equivalent to $0.20 to $0.29 per gallon of gasoline. For 12-percent efficient modules costing $0.40 per Wp, an oxygen credit of this size would cut costs for electrolytic hydrogen from $14.0 to $12.5 per GJ. As the PV hydrogen industry develops, however, the market for oxygen would become saturated and its market price would fall. Nevertheless, byproduct oxygen sales could help establish the embryonic PV hydrogen industry.

Hydrogen Storage, Compression and Transmission

If hydrogen is produced from PV electricity, the average output of the electrolysis plant is only one fourth of the peak output, since the sun shines only in the daytime. If the produced hydrogen is to be transported long distances in costly pipelines, it would make sense to run those pipelines full, so as to maximize use of the pipeline capital investment. Fortunately, storing some of the hydrogen produced in the day so as to level out the

quantity of hydrogen delivered to the pipeline is not costly. A number of proven methods are available for storing large quantities of hydrogen.[98] For the Southwest, an attractive storage option might be to use depleted natural gas wells. For this option, the cost of the storage needed to levelize the hydrogen input to the pipeline would be about $0.20 per GJ, equivalent to about $0.026 per gallon of gasoline.

For long-distance transport in pipes, hydrogen would have to be pressurized enough to overcome frictional drag. One option would be to pressurize the gas moderately at the pipeline inlet and provide a number of booster compressors along the way. The alternative would be to provide a very high pressure at the pipeline inlet. For two reasons the latter strategy appears preferable. First, one large compressor tends to be less costly than an equivalent number of smaller units. Second, the compressor work at the pipeline inlet could be provided by the low-cost power from the PV array, whereas most booster stations would have to rely on more costly electricity from other sources.

For the case where gas is transported 1000 miles without the aid of booster compressors, with inlet and exit pipeline pressures of 1000 psia (68 atmospheres) and 300 psia (20 atmospheres), respectively, the total cost of compression would be $1.2 to $1.5 per GJ for PV electricity costing $0.020 to $0.035 per kWhDC.[99] The contribution of compressor work to the cost of the produced hydrogen is equivalent to gasoline costing $0.16 to $0.20 per gallon—a non-trivial cost, but one which is nevertheless only 11 to 14 percent of the total production cost.

While there are not significant economies of scale for electrolytic hydrogen production at sizes greater than 2 MW, there are major scale economies for long-distance hydrogen transport. Consider the cost of transporting in a single pipe for 1000 miles hydrogen equivalent to about 1 percent of total U.S. oil and gas use (16,000 MW). For inlet and exit gas pressures of 1000 and 300 psia, respectively, the required pipe would have to be about 1.6 meters in diameter (63 inches) and installed would cost more than one thousand dollars per linear meter! Even though 1000 miles of such a pipeline would cost $1.7 billion, the cost per GJ of delivered hydrogen would be a relatively modest $0.35 per GJ, equivalent to $0.05 per gallon of gasoline.[100]

Some researchers have suggested that existing natural gas pipelines might be used for hydrogen transmission. Although natural gas pipelines are not ideal for hydrogen transmission,* a relatively small cost penalty would be incurred for the mismatch.[102] As shown in Figure 10, the natural gas pipeline grid in the United States runs primarily from gas wells in the Gulf states and the Southwest—areas with plentiful sunshine (Table 5), which would be promising sites for PV hydrogen production. At the very least, hydrogen pipelines could be built along existing natural gas pipeline rights of way. As natural gas production in the Gulf states and Southwest falls off, an increasing percentage of PV hydrogen could be mixed in. A similar strategy has already been used by some U.S. utilities, which routinely blend hydrogen-rich refinery off-gases with natural gas. In Honolulu, manufactured gas typically contains 10 percent hydrogen by volume.[103]

------

*Problems of hydrogen embrittlement and diffusion, which could be problematic in converting some kinds of gas-handling equipment to hydrogen, would not be serious problems for pipelines.[101] Embrittlement would not be a serious problem under the modest temperature and pressure conditions in pipelines. Compressors, valves, and other flow-modifying parts for which diffusion might be problematic would typically have to be replaced anyway when a pipeline is converted to hydrogen to accommodate the different fluid characteristics of hydrogen.

Hydrogen pipelines could be built along existing natural gas pipeline rights of way. As natural gas production in the Gulf states and Southwest falls off, an increasing percentage of PV hydrogen could be mixed in.

The Cost of Delivered Hydrogen

When the costs of electrolysis at atmospheric pressure, storage, compression, and pipeline transport for 1000 miles are considered together, the total cost of PV hydrogen at the end of the pipeline can be calculated as a function of the PV DC electricity cost. (See Figure 7.)

The delivered cost of PV hydrogen for a particular service depends on the pressure required and local distribution and delivery costs. As shown in Chapters 6 and 7 (where the economics of PV hydrogen for transport and residential heating are discussed), if PV modules become 12- to 18-percent efficient and their costs come down to $0.2 to $0.4 per Wp, delivered hydrogen could cost $12.9 to $18.0 per GJ for transport use and $12.5 to $17.6 per GJ for residential use. These costs are not low, compared to present energy prices. Indeed, the costs of hydrogen for transport would equal those of gasoline costing $1.68 to $2.35 per gallon. Yet, as is shown in Chapter 5, these costs are comparable to those for the various synthetic fuels that are being considered as alternatives to liquid and gaseous fossil fuels.

Figure 10: The U.S. Natural Gas Pipeline System.

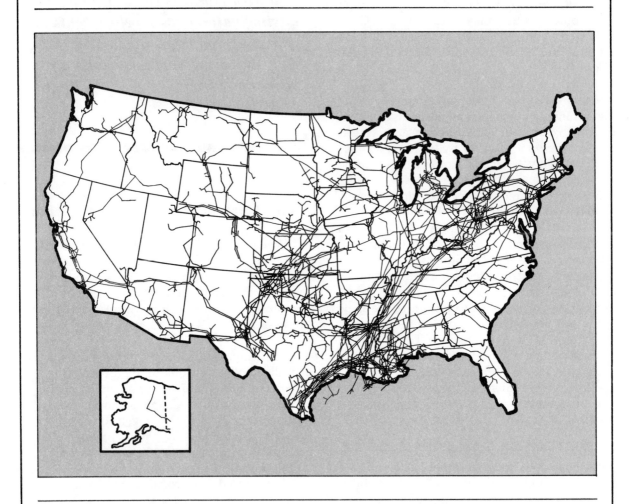

Source: Federal Energy Regulatory Commission, based on reports and maps filed with the Commission as of July 31, 1988.

40

V. How PV Hydrogen Compares to Other Synthetic Fuels

The ultimate role of PV hydrogen in the energy economy will depend on how PV hydrogen compares to the alternative fuels with which it would be competing at the turn of the century and the decades immediately thereafter.

High on the list of criteria by which PV hydrogen will be judged are environmental considerations. Compared to conventional fossil fuels, hydrogen offers significant environmental benefits. But PV hydrogen is not the only fuel that offers these advantages. For example, electrolytic hydrogen from nuclear-, hydro-, or wind-power sources would also result in negligible local, regional, and global air pollution, and synthetic fuels derived from renewable biomass sources (including ethyl alcohol from sugar cane or corn and methyl alcohol produced from a wide range of cellulosic feedstocks, including wood and urban refuse[104]) would emit no sulfur oxides and would not contribute to the atmospheric build-up of carbon dioxide. Because PV hydrogen is not the only low-polluting fuel, it must offer marked advantages relative to these alternative fuels, its direct economic costs must be acceptable, and it should not pose other serious environmental or safety problems if it is to become the fuel of choice in the future.

PV and Nuclear Electrolytic Hydrogen

Widespread concern about the greenhouse problem is leading to calls to reconsider nuclear power. For example, participants in the 1988 World Conference on The Changing Atmosphere recommended revisiting the nuclear power option, saying that:[105]

"If the problems of safety, waste, and nuclear arms proliferation can be solved, nuclear power could have a role to play in lowering CO_2 emissions."

One of the unsolved problems mentioned here, nuclear safety, is being given focused attention in some quarters of the nuclear community.[106] In principle, nuclear power plants can be designed to be quite safe. And although satisfactory technical and institutional solutions to the waste-disposal problem have not yet been demonstrated, there are no fundamental technical reasons why such solutions cannot be found.

But the third problem—the nuclear weapons connection to nuclear power—is more troubling. Inherent in nuclear technology is the fact that the chain-reacting materials that produce energy inside a reactor—the uranium isotopes U-235 and U-233 and plutonium—can also be used to make nuclear explosives. Current nuclear reactors use natural or slightly enriched uranium, which cannot be used without further enrichment to make nuclear weapons. But plutonium is an inevitable byproduct of the production of energy in these reactors. A one gigawatt (GW) light water reactor (LWR), the dominant type in most of the world, discharges approximately 200 kg of plutonium annually in its spent fuel—enough for more than 20 nuclear weapons.

In a nuclear industry large enough to significantly reduce the greenhouse problem, the potential for diversion of weapons-usable materials would be daunting. Even replacing one fourth of the present global level of fossil-fuel use (substituting nuclear electricity for coal-based electricity and hydrogen derived from nuclear electricity for oil and natural gas) could require increasing global installed nuclear capacity from the 1986 level of 300 GW to 3000 GW, equivalent to about twice the present level of global electricity generation from all sources. With current reactor technology, a nuclear capacity of this magnitude would produce more than 500,000 kilograms of plutonium per year.

This is bad enough. But concerns about uranium supply at such a high level of nuclear power development could lead to "plutonium recycling" to improve uranium efficiency. With recycling, the plutonium discharged in the spent fuel of reactors is separated from the highly radioactive fission products in so-called "reprocessing" plants and then recycled in the fresh fuel of reactors. Indeed, the initial stages of such recycling are already underway in Europe and Japan. Under current plans for this second-generation nuclear power technology, more than 25,000 kg of plutonium will be circulating in trucks, trains, ships and planes in worldwide commerce in hundreds of shipments annually by the year 2000.[107]

A 3000 GW nuclear system based on plutonium-breeder reactors would place into global nuclear commerce each year approximately five million kilograms of separated plutonium, enough to make a half million nuclear weapons.[108] It is difficult to imagine human institutions capable of safeguarding such plutonium flows against occasional diversions of significant quantities to nuclear weapons purposes.[109]

Such a large role for nuclear power is unlikely, however, unless it is dictated by political forces since the economics of nuclear power are not compelling. At today's nuclear power costs in the United States, electrolytic hydrogen from nuclear power would cost almost twice as much as the high end of the costs projected here for PV hydrogen in the year 2000. And even if the cost reduction goals for a reborn nuclear industry were met—involving nearly 50 percent reductions in both capital and O&M costs—the cost of nuclear electrolytic hydrogen would be no lower than the upper end of the range of costs targeted for PV hydrogen in the year 2000.* *(See Table 8.)*

PV Hydrogen and Electrolytic Hydrogen from Other Renewable Sources

Low-cost hydrogen can be produced at many existing hydroelectric sites, but natural limits on potential global hydroelectric resources will keep hydrogen produced this way from having much of a global impact in replacing fossil fuels.[110] Most new hydroelectric supplies cost much more than the low-cost supplies already developed; by the early part of the next century, they will probably cost much more than will PV DC electricity based on a-Si solar technology. *(See Table 8.)*

Electrolytic hydrogen could also be produced from wind power. But good wind sites are also quite limited globally and confined largely to the already industrialized world,[111] and hydro-

---

*PV hydrogen would probably be cheaper than nuclear hydrogen even if some of the cost goals for a-Si technology were not met. For example, if the balance of system were to cost $50 per m² ($50 per m² is the DOE goal for the year 2000 and has already been achieved in experimental systems in Europe) instead of $33 per m² or if the a-Si modules were to last not 30 but 15 years, the cost of PV hydrogen would still be in the range $10 to $18 per GJ, compared to $14 to $24 per GJ for nuclear hydrogen. *(See Table 9.)*

Table 8. Estimated Costs for Alternative Sources of Electrolytic Hydrogen

	PV[a]		Nuclear[b]		Hydropower[c] (Global Averages)		Wind[d]	
	n=12% $0.4/W_p	n=18% $0.2/W_p	Current	Target	2000L	2000H	Low	High
Electricity Generation								
System Size (MW)	10	10	1100	1100	–	–	(40 × 2.5)	
Capital Cost ($/kW)	992	564	2970	1620	3260	4000	1340	1580
Plant Life (years)	30	30	30	30	50	50	30	30
Capacity Factor	0.271	0.271	0.566	0.65	0.47	0.47	0.35	0.35
O & M (mills/kWh)	1.9	1.3	12.0	6.5	2.9	2.9	8.7	10.3
Fuel (mills/kWh)	–	–	7.5	7.5	–	–	–	–
Electricity Production								
Cost (mills/kWh)	34.7	19.8	66.5	36.3	55.4	70.2	43.0	50.8
Hydrogen Production Cost[e] ($/GJ)								
Electricity	11.47	6.55	22.91	12.51	19.01	24.18	14.83	17.48
Electrolyzer	2.52	2.52	1.47	1.27	1.76	1.76	2.36	2.36
Total	14.0	9.10	24.4	13.8	20.8	25.9	17.2	19.8

a. See Note 94 and Table 9 (for a sensitivity analysis of PV hydrogen costs.)
b. The capital costs indicated for nuclear power are estimates made by the Electric Power Research Institute (EPRI, *Technical Assessment Guide 1: Electricity Supply, 1986*). The higher value is EPRI's estimate of the cost of a plant that would be ordered in the United States at present. The lower value is EPRI's target for "improved conditions" in the United States—resulting from higher construction labor productivity, a shorter construction period, a streamlined licensing process, etc. The "current" capacity factor (56.6 percent) is the actual average for U.S. nuclear plants from 1983 to 1987; for "target" conditions it is assumed that this increases to 65 percent. The fuel cost of 0.75 cents per kilowatt hour and the current O&M cost are actual average values for nuclear plants in 1986 [Energy Information Administration, "Historical Plant Costs and Annual Production Expenses for Selected Electric Plants, 1986," DOE/EIA-0455(86), May 27, 1988]. The "target" O&M cost is from EPRI (EPRI, 1986).
c. The hydro capital costs and capacity factors are global average estimates for the year 2000 (H.K. Schneider and W. Schulz, *Investment Requirements of the World Energy Industries 1980–2000*, World Energy Conference, 1987). The hydro O&M costs are EPRI estimates for the indicated capacity factor (EPRI, 1986).
d. For mass-produced 2.5-MW wind turbines configured to produce 100 MW (J.I. Lerner, "A Status Report on Wind Farm Energy Commercialization in the United States, with Emphasis on California," paper presented at the 4th International Solar Forum, Berlin, FRG, October 6–9, 1982). The O&M cost is assumed to be 2 percent of the initial capital cost per year.
e. For non-PV electricity sources, unit electrolyzer capital costs are assumed to be 25 percent higher because of the rectifier, which is assumed to be 96-percent efficient.

gen derived from wind is likely to cost considerably more than hydrogen derived from PV sources. *(See Table 8.)*

PV Hydrogen and Biomass-Derived Fluid Fuels

As the comparison of PV hydrogen and biomass-derived fuels in Table 9 shows, the cost estimates for ethanol derived from sugar cane and corn ($8 to $14 per GJ) are in the same range as the cost targets set forth in this study for PV hydrogen ($9 to $14 per GJ). Indeed, biomass converted to modern energy carriers has an important role to play in the world's energy economy, especially in developing countries.[112]

However, two fundamental constraints limit bioenergy development over the long term: large land-area requirements arising from the low efficiency of photosynthesis, and large water requirements. While solar energy can be converted to hydrogen energy using solar cells at an overall efficiency of 8 to 13 percent, the maximum efficiency of photosynthesis is probably about 5.5 percent. In practice, photosynthetic efficiency is closer to 1 to 4 percent under ideal growing conditions and only one-sixth to one-half percent under average conditions.[113] This low efficiency is compounded by another problem: the conversion of raw biomass to a high quality fluid fuel typically involves energy losses in the range of 30 to 50 percent. As a result, at high levels of bioenergy utilization, land requirements become formidable—much greater than the land required for PV hydrogen.

To illustrate the problem, consider the land-use requirements for fluid fuels derived at 65 percent conversion efficiency from biomass grown on energy farms or plantations with a biomass productivity of 40 tonnes of dry biomass per hectare per year. While these values for the conversion efficiency and biomass productivity are both considered relatively high, the overall efficiency of converting sun-

light into fluid fuels this way is only about 0.75 percent for typical solar insolation values. Such a low overall efficiency implies that in the United States, for example, providing bioenergy in an amount equivalent to the present level of fossil fuel use would require a land area more than half that presently used as cropland or forests. *(See Table 10.)*

The land requirements for PV hydrogen production would be less than one tenth of those for bioenergy production, so even densely populated countries would have the space to produce PV hydrogen.

In contrast, the land requirements for PV hydrogen production would be less than one tenth of those for bioenergy production, so even densely populated countries would have the space to produce PV hydrogen. *(See Table 10.)* Producing hydrogen equivalent to total U.S. oil use would require a collector field in the Southwest of some 64,000 square kilometers (24,000 square miles)—roughly 0.5 percent of total U.S. land area or about 7 percent of the desert area in the United States. *(See Figure 11.)* Hydrogen equivalent to total world fossil fuel consumption could be produced on a mere 530,000 square kilometers (205,000 square miles) of arid lands, less than 2 percent of the world's deserts. *(See Tables 10 and 11.)*

Lack of adequate water supplies further constrains bioenergy development. Biomass production requires some 300 to 700 tonnes of water (for photosynthesis and transpiration) per tonne of dry biomass produced. Assuming that fluid fuels are produced from biomass at an average efficiency of 65 percent, the water requirements become some 25,000 to 60,000 liters per GJ of produced biofuel.

Table 9. Estimated Production Costs for Synthetic Liquid and Gaseous Fuels[a]

Fuel	Plant Size (1000 GJ/day)	Synfuel Plant Installed Cost ($10^6)	($/W_p)	Fuel Production Cost[b] ($/GJ)	($/gal)[e]
Gasoline Derived from Coal	176	5107	2.26	16.2	2.11
Methanol Derived from Coal	322	3860	0.93	7.9	1.03
	32.2	567	1.37	10.0	1.30
Ethanol Derived from					
Sugar cane (Brazil)	2.68	7.7	0.22	8.0	1.04
Corn (US)	13	95.2	0.57	14.4	1.88
Synthetic Gas Derived from Coal[c]					
High heating value gas	264	1820	0.54	5.7	0.74
	88	756	0.67	6.4	0.83
	26.4	341	1.00	7.9	1.03
Intermediate heating value gas	264	822	0.24	4.0	0.52
PV Hydrogen (Southwestern US)					
BOS = $33/m², PV System Lifetime = 30 years					
n=18%, $0.2/W_p	0.183[d]	7.0	0.89	9.1	1.19
n=12%, $0.4/W_p		10.9	1.40	14.0	1.83
BOS = $50/m², PV System Lifetime = 30 years					
n=18%, $0.2/W_p	0.183[d]	8.2	1.06	10.6	1.38
n=12%, $0.4/W_p		12.9	1.65	16.3	2.12
BOS = $33/m², PV System Lifetime = 15 years					
n=18%, $0.2/W_p	0.183[d]	7.0	0.89	10.4	1.35
n=12%, $0.4/W_p		10.9	1.40	18.2	2.36

a. Cost estimates were derived using a self-consistent set of assumptions to make comparison of technologies meaningful. (*See Notes 92, 94, 114.*)

b. The production cost includes plant capital, feedstock and operation and maintenance costs, but not the costs of transmission or storage.

c. For the Lurgi, dry-ash process with Western U.S. coal.

d. This corresponds to a 9.3 MW_p PV array coupled to a 10 MW_p, 84 percent efficient unipolar electrolyzer, for which the coupling efficiency is 93 percent. The peak PV hydrogen output is 7.8 MW, and the average output is 183 GJ/day for a system located in the southwestern United States where the average insolation is 271 Watts per square meter. There should be no significant economies of scale for PV-powered electrolyzers above about 5-10 MW.

e. The cost of energy in dollars per gallon of gasoline-equivalent.

In contrast, the consumptive water requirements for electrolysis are a modest 63 liters per GJ of hydrogen produced (assuming the water used to cool the electrolyzer is recirculated). In producing an amount of hydrogen equivalent in energy to one liter of gasoline, only two liters of water are consumed. For comparison, per capita water use in the U.S. is more than 100 times the per capita level of petroleum consumption (in volumetric terms).[115] Thus, the

Table 10. Land Use Requirements for PV Hydrogen, Biomass and Coal Synfuels

Fossil Fuel Displaced at the Present U.S. Average Per Capita Consumption Rate	Fossil Fuel Displaced/Year (Gigajoules/person)	Land Required (hectares/person)		
		w/PV H$_2$[a]	w/Biomass[b]	w/Coal Synfuels[c]
	287	0.053	0.61	0.018

U.S. Land Area Per Capita (hectares/person) for[d]:

Forests and Woodland	1.1
Cropland	0.8
Permanent Pasture	1.0

Average Land Area per Capita[d] in:

World	2.62
Africa	5.02
Rwanda	0.38
North America	5.18
El Salvador	0.35
Asia	0.92
Bangladesh	0.13
Europe	0.96
Netherlands	0.23
South America	6.29
Ecuador	2.79

Fossil Fuel Displaced for the World at the Present Consumption Rate	Fossil Fuel Displaced/Year (Exajoules)	Land Required (million hectares)		
		w/PV H$_2$[a]	w/Biomass[b]	w/Coal Synfuels[c]
	283	53	600	18

Total World Land (million hectares) in:

Forests and Woodland[d]	4090
Cropland[d]	1470
Permanent Pasture[d]	1350
Deserts[e]	3140

a. For PV electricity produced in 15-percent efficient PV systems (with tilted collectors requiring a ground area twice as large) and converted to hydrogen at 84 percent efficiency. Average insolation on tilted collectors is 271 Watts/square meter (250 Watts/square meter on the ground.)

b. For biomass grown @ 40 tonnes of dry biomass (@ 18 GJ/tonne) per hectare per year and converted to fluid fuels at 65 percent efficiency, on average.

c. For surface-mined coal. (See Note 117.)

d. World Resources Institute, *World Resources 1987*, Basic Books, Washington, DC, 1987.

e. M.P. Petrov, *Deserts of the World*, John Wiley & Sons, 1976.

Figure 11: Land Requirements for PV Hydrogen Production in the United States.

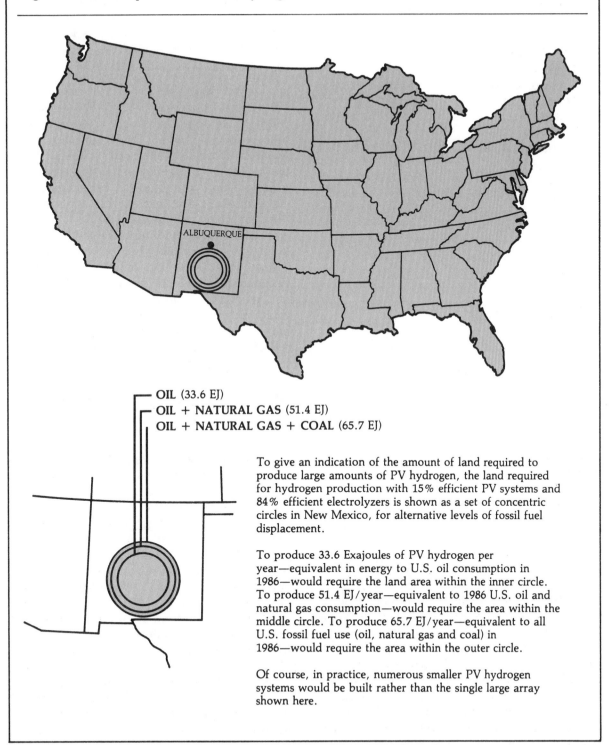

ALBUQUERQUE

— OIL (33.6 EJ)
— OIL + NATURAL GAS (51.4 EJ)
OIL + NATURAL GAS + COAL (65.7 EJ)

To give an indication of the amount of land required to produce large amounts of PV hydrogen, the land required for hydrogen production with 15% efficient PV systems and 84% efficient electrolyzers is shown as a set of concentric circles in New Mexico, for alternative levels of fossil fuel displacement.

To produce 33.6 Exajoules of PV hydrogen per year—equivalent in energy to U.S. oil consumption in 1986—would require the land area within the inner circle. To produce 51.4 EJ/year—equivalent to 1986 U.S. oil and natural gas consumption—would require the area within the middle circle. To produce 65.7 EJ/year—equivalent to all U.S. fossil fuel use (oil, natural gas and coal) in 1986—would require the area within the outer circle.

Of course, in practice, numerous smaller PV hydrogen systems would be built rather than the single large array shown here.

Table 11. Desert Areas by Continent (million hectares)		
North America		190
Temperate Zone	60	
Subtropical Zone	90	
Tropical Zone	40	
South America		180
Temperate Zone	50	
Subtropical Zone	50	
Tropical Zone	80	
Eurasia		1430
Temperate Zone	590	
Subtropical Zone	470	
Tropical Zone	370	
Africa		1000
Subtropical Zone	110	
Tropical Zone	890	
Australia		340
Subtropical Zone	20	
Tropical Zone	320	
TOTAL		3140

Source: M.P. Petrov, *Deserts of the World*, John Wiley & Sons, 1976.

water requirement for enough hydrogen to replace current petroleum use would be a "drop in the bucket," adding only about 2 percent to the average per capita water use in the United States.

Water requirements are also quite modest compared to precipitation levels in areas where PV hydrogen systems might be located. A typical average insolation value for hot arid regions is some 270 Watts per square meter. If the land area requirements were twice the PV collector area (to prevent significant self-shading), the water requirements for hydrogen production would be some 2 to 3 centimeters per year (See Table 12)—a small fraction of total precipitation even in very arid regions. Consequently, PV hydrogen can be produced even in deserts. For example, even though El Paso is in one of the driest areas in the United States—with annual rainfall amounting to only 20 centimeters (8 inches)—the water required for electrolysis is only 12 to 17 percent of total precipitation falling on the PV collector field.[116]

Biomass does offer something that PV hydrogen cannot, however: a source of carbon that is not derived from fossil fuel. Carbon-based fuels will be preferable to hydrogen in some applications—notably, where liquid fuels are required. While hydrogen can be liquefied, the use of liquid hydrogen as a fuel is beset by many practical complications, because it is a cryogenic fuel. (See Chapter 6.) In contrast, there are many carbon-based liquid fuels that are easily stored under ambient conditions. Biomass may ultimately become very valued because it can serve as a feedstock for convenient liquid fuels. In this regard an important synergism could develop: PV hydrogen could become a source of hydrogen for biomass-derived liquid fuels, thereby significantly reducing the amount of land and water needed to produce a given amount of liquid fuel.

PV Hydrogen and Synthetic Fossil Fuels

Perhaps the most troubling feature of coal-based synthetic fossil fuels is that a shift to these fuels would speed up the greenhouse warming because the carbon dioxide emissions per unit of useful energy derived would be much higher than with conventional oil and gas. Synthetics from coal would release almost twice as much carbon dioxide per unit of energy consumed as oil and almost three times as much as natural gas. (See Figure 3.) Carbon dioxide emissions from a natural gas-based fuel such as methanol would be comparable to those from gasoline, but since natural gas supplies are limited, widespread use of such a fuel would inevitably lead to a switch to more abundant coal feedstocks. From the standpoint of controlling greenhouse gas emissions, PV hydrogen is clearly far more attractive than fossil synthetics.

Table 12. Water Requirements for PV-Hydrogen and Biomass Energy Systems

	Useful Energy Production Rate (GJ/m²/year)	Water Requirements (cm of rainfall/year)
Efficiency of PV modules (%)[a]		
10	0.35	2.3
15	0.53	3.4
Biomass Productivity[b] (tonnes/hectare/year)		
10	0.012	30– 70
20	0.023	60–140
30	0.035	90–210
40	0.046	120–280

a. *See Note 116.*
b. Biomass production requires for photosynthesis and transpiration some 300 to 700 tonnes of water for each tonne of dry biomass produced. Assuming that fluid fuels are produced from biomass at an average efficiency of 65 percent (and that the heating value of dry biomass is 18 GJ per tonne), the water requirements become 25,000 to 60,000 liters per GJ of produced biofuel.

From a land-use perspective, fossil synthetics might seem greatly preferable to PV hydrogen. But while land use requirements for fossil synthetics are probably less than for PV hydrogen, the advantage is not decisive: the amount of land required to produce synthetic liquid fuels equivalent to current U.S. oil use from the estimated 86 billion tonnes of stripminable coal reserves would average about 8500 square miles; the amount of land required for the same amount of PV hydrogen is only about three times as large.[117] *(See Table 10.)*

From a narrow economic perspective, PV hydrogen would probably not have a decisive edge over fossil synthetic fuels. *(See Table 9.)* PV hydrogen would probably be approximately competitive with synthetic liquids from coal (projected to cost $8 to $16 per GJ), but more expensive than synthetic gases from coal (projected to cost $4 to $8 per GJ).

The production cost comparisons shown in Table 9, based on detailed systems analysis but not on actual field experience, may prove to be biased in favor of synthetic fossil fuels because the low cost estimates reflect the assumption that synfuels will be made in huge facilities. To achieve favorable production economics for methanol from coal, for example, a facility costing nearly $4 billion and having a methanol production capacity of some 3700 MW (322 thousand GJ per day) must be constructed. For production at a rate of 37 MW (3.22 thousand GJ per day), the cost of the produced methanol would be by 70 percent higher. In contrast, because PV arrays and electrolyzers are modular technologies, PV hydrogen production costs could be low in very small facilities.

Since PV systems and electrolyzers are modular they can be largely factory-built, and manufacturers can exploit the economies of mass production, while synthetic fossil fuel plants will require extensive field construction, which is inherently more costly. The importance of this difference, though not generally appreciated, was perceptively articulated by John Fisher in 1974 in an analysis of the escalation in nuclear power costs in the decade leading up to the first oil crisis:[118]

"When measured in constant dollars per kilowatt of capacity, the cost of constructing a nuclear power plant increased by perhaps 50 percent in the past decade... When power plant costs rise an explanation is required, as we expect all power plant costs to decline through the economies of scale and new technology. The environmental movement was responsible for part of the rise in nuclear plant costs, by causing various procedural delays and by requiring additional expensive safeguards to protect against hypothetical accidents. But there appears to be another cause for increasing construction costs, associated with a growing portion of high-cost field construction and a shrinking portion of low-cost factory construction for the very large power plants now being built...the costs associated with a shift to field from factory can more than offset anticipated economies of scale...."

Fisher pointed out that, historically, as the electric utility industry and plant capacity doubled every decade, factory capacity also doubled, as did field construction at each site. Manufacturing and construction costs per kW declined in the factory and in the field since both increased their scale of operations. As long as both activities grew in proportion, the economies of scale produced similar cost reductions in each, and therefore an overall cost reduction, even though the unit cost of field construction was always higher than the unit

cost of factory construction. This pattern held until plant size reached about 200 MW. Then, because design engineers felt that scale economies would be much more important for nuclear than for fossil fuel plants, nuclear power plant capacities were built in sizes of the order of 1000 MW—shifting a greater portion of the construction from the factory to the field. Fisher's important insight is that the widely touted economies of scale in power plant construction are illusory because field construction is inherently more costly than factory construction, and because with field construction it is never possible to get very far down the "learning curve," as is possible with factory production.

Hydrogen Safety

Hydrogen is clearly preferable to other fluid fuels from an environmental perspective, but is it safe? Everyone remembers the Hindenburg disaster, and this event often conjures up images of hydrogen as an inherently unsafe fuel.

The very properties that make fluid fuels so useful—the capacity to store a large amount of readily accessible energy in a small volume—also make them potentially dangerous. Despite the widespread perception that hydrogen is much more dangerous than natural gas or gasoline, this concern is not borne out by the fuel's physical properties or by many years of industrial and residential experience with hydrogen. *(See Box 3.)* Studies of the relative safety of hydrogen, methane (the dominant constituent of natural gas), and gasoline have concluded that no one fuel is inherently safer than the others in every respect. Each fuel demands its own set of precautions, but all three fuels can be (and have been) used safely.

Box 3. Is Hydrogen Safe?

What are the hazards of using hydrogen as compared to those of using such common fuels as natural gas and gasoline?

A Comparison of the Physical Properties of Hydrogen, Gasoline and Methane

Some safety-related physical properties of hydrogen, methane (natural gas is typically about 96 percent methane) and gasoline are listed in Table B.3.1. A few significant properties that are most often mentioned in discussions of hydrogen safety include:

Limits of flammability (detonability) in air. The limits of flammability (detonability) indicate the range of mixtures of fuel in air which will sustain a fire (sustain an explosion). Hydrogen will burn (detonate) over a much wider range of concentrations than either methane or gasoline.

While hydrogen's wide range of flammability would seem to be a serious concern, the lower flammability limit would be the more relevant parameter, if the fuel concentration built up gradually through a leak, as might occur in residential or indoor industrial settings. This is true because sources of ignition are often present in residential or industrial environments, and a fire would be likely as soon as the fuel-to-air concentration reached the lower flammability limit. Hydrogen's lower flammability limit (4 percent by volume) is not much less than that for methane (5.3 percent) and is higher than that for gasoline (1 percent); hydrogen's lower detonability limit is 18.3 percent, significantly higher than that of methane (6.3 percent) or gasoline (1.1 percent). It is obviously important to detect leaks and provide adequate ventilation for all three fuels, when they are used in enclosed spaces.

The range of flammability (detonability) could be a concern in hydrogen transmission and distribution systems. It might be necessary to purge a hydrogen pipeline with an inert gas (for example, nitrogen) prior to commissioning or after a maintenance shutdown, so that flammable (or detonable) hydrogen-air mixtures don't form. This is the current industrial practice with hydrogen-handling systems. During normal operation, of course, a hydrogen pipeline or distribution system would be pressurized, and air could not leak in.

Minimum energy for ignition. Hydrogen has a significantly lower ignition energy than gasoline or methane (e.g. it ignites more easily than either methane or gasoline). However, the ignition energy for all three fuels is extremely low, and weak thermal sources of ignition (sparks from static electricity, open flames from matches, hot surfaces) would be more than sufficient to ignite flammable mixtures of methane, gasoline or hydrogen.

Buoyant velocity/diffusion velocity in air. Buoyant velocity refers to the rate at which fuels rise in air, diffusion velocity to the rate at which they diffuse through air. Both quantities reflect how quickly fuel vapors would disperse after a leak. Hydrogen is the lightest element, and it rises and disperses very quickly. Methane is also buoyant and disperses quickly, though less so than hydrogen. Gasoline vapors are heavier than air and can linger near a leak even outside. There should be little problem with hydrogen or methane leaks outside (from storage, pipelines or an external automotive tank), since these fuels would tend to disperse before a flammable mixture could build up. This is not true with gasoline: fumes could reach a hazardous concentration more easily. For indoor use, adequate ventilation

Table B.3.1. Safety-Related Physical Properties of Hydrogen, Methane and Gasoline

	HYDROGEN	METHANE	GASOLINE
		(96% of natural gas)	
Limits of flammability in air (% volume)	4.0–75.0	5.3–15.0	1.0–7.6
Limits of detonability in air (% volume)	18.3–59.0	6.3–13.5	1.1–3.3
Minimum energy for ignition in air (millijoules)	0.02	0.29	0.24
Diffusion velocity in air (meters per second)	2.0	0.51	0.17
Buoyant velocity in air (meters per second)	1.2–9.0	0.8–6.0	non-buoyant
Leak rate in air (relative to methane)	2.8	1	1.7–3.6 (gasoline vapors)
Toxicity	non-toxic	non-toxic	toxic in concentrations > 500 ppm

SAFETY ASPECT	RELEVANT PHYSICAL CHARACTERISTICS	IMPLICATIONS
Fire/Explosion Hazard Leak Rate	Density, absolute viscosity	Hydrogen has a leak rate 2.8 × methane 0.6–7.3 × gasoline
Ignitability	Minimum ignition energy, Flammability/Detonability limits, Buoyant velocity, Diffusion velocity.	All three fuels ignite very easily. Ignition hazard persists longest w/gasoline then with CH_4, H_2.
Physiological Hazard		Gasoline toxic, H_2 and CH_4 non-toxic

a. Adapted from J. Hord, "Is Hydrogen a Safe Fuel?" *International Journal of Hydrogen Energy*, v. 3, pp. 157–176, 1978. Adapted from G.D. Brewer, "Some Environmental and Safety Aspects of Using Hydrogen as a Fuel," *International Journal of Hydrogen Energy*, v. 3, pp. 461–474, 1978.

Box 3. (cont.)

(through roof vents) would assure that buoyant hydrogen or methane would not reach flammable concentrations.

Leak rate. Hydrogen gas would have a volumetric leak rate (for example, from a cracked weld or damaged seal) about three times that of methane gas, but less than that of gasoline vapors. Hydrogen also disperses more quickly than methane or gasoline, so that it would be more difficult to reach dangerous concentrations of fuel in air. For indoor use, roof vents would be desirable to assure that buoyant hydrogen (or methane) would be eliminated from a closed space in case of a leak. As with natural gas, an odorant would probably be added to hydrogen to aid detection of leaks. Experience in industrial and residential settings has shown that, with appropriate precautions, all three fuels can be safely handled, distributed and stored, with minimal leakage.

Materials embrittlement. At high temperatures and pressures (such as those found in hydrogen handling equipment in oil refineries) hydrogen can cause potentially dangerous embrittlement of metal vessels and pipes. However, at the lower temperatures and pressures found in a hydrogen energy system, embrittlement should not be a problem.

Toxicity. Hydrogen and methane are nontoxic, while gasoline and gasoline vapors are toxic. All three gases are asphyxiants at high concentrations.

Residential Experience with Hydrogen

Since the 19th century, hydrogen-rich "town gas" manufactured from coal or wastes has been widely used as a home heating fuel. Many utilities in the United States switched from town gas to natural gas as recently as the 1940s and 50s, when long-distance pipelines were built linking gas fields in the Gulf states and the Southwest to the Northeast and Midwest. While it is now rare in the U.S., town gas is still common in many regions of the world where natural gas is expensive or unavailable. Town gas is typically about half hydrogen and half carbon monoxide (with traces of methane and other combustible gases). The flammability and combustion properties of carbon monoxide are similar to those of hydrogen, so that the fire and explosion hazards with town gas would be similar to those for pure hydrogen. The fire hazards associated with residential use of town gas appear to have been acceptable, suggesting that hydrogen too would be acceptable for residential applications.

Industrial Experience with Hydrogen

Hydrogen has been used routinely in process industries for decades, and safe storage and handling techniques have been developed. Hydrogen pipelines several hundred kilometers in length have been operated in Germany, the United States and England, with no undue problems, and utilities have used hydrogen for generator cooling for fifty years. The overall conclusion of past studies is that, with proper handling, safety should not be a problem in industrial or utility settings.

Conclusions

Chemical fuels such as natural gas, gasoline and hydrogen are potentially hazardous and must be treated with respect. No one fuel is inherently safer than others in all situations. The evidence indicates that all three can be produced, stored, transmitted and

Box 3. (cont.)

handled safely for industrial, residential and transport use.

For an excellent and extensive discussion of hydrogen safety issues the reader is referred to the paper by J. Hord.

Sources: J. Hord, ''Is Hydrogen a Safe Fuel?,'' *International Journal of Hydrogen Energy,* v. 3, pp. 157–176, 1978; G.D. Brewer, ''Some Environmental and Safety Aspects of Using Hydrogen as a Fuel,'' *International Journal of Hydrogen Energy,* v. 3, pp. 461–474, 1978; W. Balthasar, ''Safety Aspects in the Use of Hydrogen in Industry and in the Home,'' in *Hydrogen: Energy Vector of the Future,* Graham and Trotman, London, 1983; H. Tamm, ''Summary of Hydrogen Safety Workshop held at the Fifth World Hydrogen Energy Conference,'' *International Journal of Hydrogen Energy,* v. 11, pp. 61–66, 1986.

VI. How PV Hydrogen Could Replace Oil

PV Hydrogen as a Fuel for Transportation

While hydrogen could supplant oil in virtually all its current end uses, transportation markets are particularly important, since they account for over half of oil use in the United States and more than one third of oil use worldwide. Widespread use of PV hydrogen as a transport fuel could greatly reduce fossil fuel use in response to the greenhouse problem, as well as reduce dependence on oil.

Widespread use of PV hydrogen as a transport fuel could greatly reduce fossil fuel use in response to the greenhouse problem, as well as reduce dependence on oil.

Air quality would also improve if PV hydrogen were substituted for oil-derived fuels in automobiles. In many U.S. cities, automobiles are responsible for the majority of NO_x and VOCs emissions.[119] They also contribute heavily to such regional pollution problems as acidification of the water and soils. Switching from gasoline to less-polluting transport fuels such as hydrogen could do much to ameliorate these problems. Indeed, pollution-control strategies mandating the use of alternative fuels in transportation have been adopted in California.

Since highly-refined transport fuels (for example, gasoline and Diesel fuel) are the most expensive fluid fuels, it might be easier for hydrogen (or other synthetic fuels) to compete on a fuel cost basis in transportation markets than in other fluid fuel markets.

And finally, hydrogen-based transport technology could probably be commercialized relatively soon. Hydrogen-powered internal combustion engines have been extensively researched and experimental hydrogen-powered cars and buses have already been built.[120] *(See Figure 12.)* The viability of a gaseous fuel infrastructure for transportation has been demonstrated in Italy, New Zealand, and Canada for vehicles operated on natural gas.[121]

PV Hydrogen as an Automotive Fuel

Considering the centrality of the automobile in U.S. transportation, the prospects for using hydrogen as a transport fuel for the automobile deserve a close look. *(See Table 13.)* If the technical challenges of using hydrogen in automobiles could be overcome, then this fuel could also be used in other transport markets (for instance, in buses and trucks), where design constraints are less challenging.

Figure 12: Hydrogen-Powered Motor Vehicles.

Both the hydrogen-powered car developed by Daimler-Benz Corporation of Stuttgart, West Germany (top) and the hydrogen-powered bus built by Billings Energy Corporation of Provo, Utah (bottom) use onboard metal hydride storage tanks (shown at rear of car).

Table 13. Oil Consumption in U.S. Transportation (1983)[a]

TYPE OF TRANSPORT	OIL CONSUMPTION	
	(Exajoules/year)	(% of Total)
Automobiles	9.34	45.4
Motorcycles	0.036	0.2
Trucks	5.95	28.9
Light trucks	2.56	12.5
Other trucks	3.39	16.5
Busses	0.15	0.7
Off-highway equipment (mining, construction, farming)	0.75	3.7
Airplanes	1.58	7.7
Ships	1.26	6.1
Rail	0.61	2.9
Pipelines	0.17	0.8
Military	0.71	3.5
Total	20.6	100

a. *Source:* C.M. Hanchey and M.C. Holcomb, *Transportation Energy Data Book*, Edition 8, Oak Ridge National Laboratory Report, ORNL-6205, November 1985.

The Storage Challenge

Unlike such liquid fuels as gasoline or methanol, hydrogen is not easily stored at ambient pressure and temperature, so special storage systems are required. For use in automobiles, hydrogen gas must be compressed for storage at high pressure (2000–3000 psia), liquefied and stored cryogenically at very low temperature, or converted to another compound (such as a metal hydride) that is more compact.[122]

Compared to other automotive fuels, hydrogen has the highest energy content per kilogram of fuel, but its energy content per unit volume is low.

Compared to other automotive fuels, hydrogen has the highest energy content per kilogram of fuel, but its energy content per unit volume is low. *(See Table 14.)* Moreover, hydrogen storage systems tend to be heavy as well as bulky. Storing enough fuel on board for a reasonable travelling range is thus a major challenge. For example, to store the energy equivalent to a 20-gallon tank of gasoline would require about 1.4 cubic meters of hydrogen gas compressed to over 160 times atmospheric pressure. For perspective, a typical station wagon has a cargo space of only 1 cubic meter, and the average sedan's trunk space is perhaps half this size.[123] (Compressed natural gas would require 0.4 cubic meters, and advanced batteries 2.7 cubic meters.[124]) And the weight of the cylinders holding the compressed hydrogen would be about 1500 kg, about the same weight as a mid-sized car.

In the past, hydrogen's low energy density led researchers to view hydrogen-fueled cars as

Table 14. Energy Storage in Automotive Fuels

FUEL	ENERGY DENSITY[a] (GJ/m³) HHV	LHV	(MJ/kg) HHV	LHV	Density (kg/m³)	Type of storage
Gasoline[b]	34.5	32.3	45.9	43.0	720	metal tank
LPG (100 psi)[c]	25.5	23.6	50.0	46.3	510	pressurized cylinder
CH_4 Gas (2400 psia)	6.16	5.51	55.5	50.0	111	pressurized cylinder
Liquefied CH_4[d] ($-161\,°K$)	23.9	21.5	43.4	39.1	550	cryogenic dewar
Methanol[b]	18.1	15.9	22.7	19.9	791	metal tank
Ethanol[b]	24.0	21.6	30.2	27.2	790	metal tank
H_2 Gas (2400 psia)	2.09	1.76	142.4	120.2	14.7	pressurized cylinder
Liquefied H_2[b] ($-253\,°K$)	9.95	8.4	142.4	120.2	71	cryogenic dewar
Metal hydride[e]	5.8	4.9	2.13	1.80	2706	hydride storage
Electricity[f,g,h]	0.42		0.144 [40 Watt-hr(Wh)/kg]		2900	lead-acid battery[f] (commercially available)
	0.48		0.191 (53 Wh/kg)		2500	Ni-Fe battery[g] (best laboratory result)
	0.90		0.324 (90 Wh/kg)		2778	Li/metal sulfide battery (EPRI projection)[h]

a. HHV is the higher heating value of the fuel; LHV, the lower heating value. Densities include the fuel weight or volume, but not the storage tank.

b. E.M. Goodger, "Liquid Fuels for Transport," *Progress in Energy and Combustion Science,* Vol. 8, p. 233, 1982.

c. Gregory and J.B. Pangborn, "Hydrogen: A Portable Fuel for Military Applications," *Proceedings 26th Annual Power Sources Conference,* (May 1974); C.M. Hanchey and M.C. Holcomb, *Transportation Energy Data Book,* Edition 8, Oak Ridge National Laboratory Report, ORNL-6205, 1985.

d. Amos Golovoy and Roberta J. Nichols, "Natural Gas Powered Vehicles," *ChemTech,* p. 359, June 1983.

e. H. Buchner, "Hydrogen Use-Transportation Fuel," *International Journal of Hydrogen Energy,* Vol. 9, p. 501, 1984; H. Buchner and R. Povel, "The Daimler-Benz Hydride Vehicle Project," *International Journal of Hydrogen Energy,* Vol. 7, pp. 259–266, 1982.

f. B. Sorensen, "Energy Storage," *Annual Review of Energy,* Vol. 9, pp. 1–29, 1984.

g. U.S. Department of Energy, *Electric and Hybrid Vehicles Program,* 10th Annual Report to Congress, DOE/CE-0179, April 1987.

h. "Electric Vehicles," *EPRI Energy Researcher,* May 1986.

short-range vehicles. However, because it is now feasible to improve automotive fuel economy to 80 to 100 miles per gallon of gasoline-equivalent fuel (see Box 4), limited range no longer poses a serious constraint on the use of hydrogen as an automotive fuel. With such high fuel economy, it becomes feasible to store enough fuel for a reasonable travelling range

Box 4. The Revolution in Automotive Fuel Economy

The oil shocks of the 1970s prompted interest in improving automotive fuel economy. As a result of Congressionally mandated fuel economy standards the average on-the-road fuel economy of the U.S. automobile fleet increased from 13.5 mpg in 1975 to 18.5 mpg by 1985, and the average fuel economy of cars purchased in the United States in 1985 was 23.1 mpg.[126]

The most energy-efficient cars on the market have far higher fuel economies. The 1986 model gasoline-fueled Chevrolet Suzuki Sprint and the 1985 model diesel-fueled Nissan Sentra and Ford Escort are four- to five-passenger cars with average on-the-road fuel economies in the range 54 to 57 mpg.

Moreover, still higher fuel economies have been demonstrated. Volkswagen, Volvo, and Renault have all built prototypes with fuel economies ranging from 66 to 70 mpg.[127] The most efficient prototype built so far is a 98 mpg four- to five-passenger car introduced by Toyota in 1985. *(See Figure 13.)* No exotic technologies are required to achieve such high fuel economies—only the use of lightweight materials, aerodynamic styling to cut air resistance, and proven high-efficiency engines and transmissions.

Nor does high fuel economy necessarily imply sluggish performance that would make highway entry or passing difficult or dangerous. The Volvo LCP 2000 prototype, with a fuel economy of 66 mpg, requires only 11 seconds to accelerate from 0 to 60 mph, compared to 17 seconds for the popular Chevrolet Cavalier with automatic transmission, which has a fuel economy of only 28 mpg.[128]

Fortunately, safety need not be compromised in building lightweight cars. One possibility for making lightweight cars safe is to make them large, offering plenty of ''crush-space'' in the event of a crash—a strategy proposed in a Battelle Memorial Laboratory design for a five- to six-passenger car that would weigh 545 kg and get 100 mpg.[129] The 69 mpg LCP 2000 built by Volvo is a lightweight (707 kg) prototype engineered to withstand 35-mph (56-km per hour) front and side impacts and 30-mph (48-km per hour) rear impacts—thereby meeting stricter standards than do cars currently sold in the United States.[130] Moreover, advanced technology still under development will make it possible to build heavy fuel-efficient cars. For example, an 80 mpg car proposed by Cummins/NASA Lewis researchers—a multifuel-capable, direct-injection, adiabatic diesel with turbocompounding—would weigh 1360 kg—approximately the average weight of new cars sold in the United States today.[131]

without sacrificing passenger room or trunk space, even for very low-density fuels.[125]

Compressed gas, metal hydride, and liquid hydrogen storage systems each have been explored in experimental vehicles. Each option has attractive features and drawbacks.

At first glance, liquid hydrogen would appear to be the most desirable option for onboard fuel storage, since it has the highest energy content per unit volume of any form of hydrogen. *(See Table 14.)* However, liquid hydrogen is a cryogenic liquid—it does not become a liquid except at very low temperatures of $-423\,°F$ ($-253\,°C$)—which leads to difficulties in distributing and storing the fuel. Since no economically feasible way of piping liquid hydrogen more than one or two miles exists, it would have to be delivered by cryo-

Toyota AXV

Introduced in late 1985, the AXV (a four- to five-passenger car) has a fuel economy of 98 mpg on the combined urban/highway test administered by the US Environmental Protection Agency. High fuel economy is achieved with the systematic application of currently available technologies: low weight (650 kg) from extensive use of plastics and aluminum, low aerodynamic drag (a drag coefficient of 0.26), a direct-injection diesel engine (the kind used in trucks), and a continuously variable transmission.

Source: Toyota press release, October 23, 1985.

genic truck or liquefied at a service station at a high cost.[132] Moreover, some liquid hydrogen evaporates and must be vented from even the best automotive-sized dewars (vacuum containers used to store cryogenic liquids). These losses can amount to 0.5 to 3 percent of the energy stored in the tank per day and could conceivably create a safety hazard in enclosed spaces such as a garage. Hydrogen losses during fuel transfer are more serious: perhaps 10 to 25 percent of the hydrogen boils off during refuelling, and other significant boil-off losses can occur at each point of the distribution chain.[133] Small onboard cryogenic storage dewars would require bulky insulation to retard evaporation, making the overall storage volume comparable to that of gas cylinders or hydride tanks. (While the cost and distribution problems associated with liquid hydrogen are formidable, there may be some benefits in engine efficiency to be gained with liquid fuel compared to gaseous hydrogen from compressed gas or hydride storage.)

Relative to other schemes, compressed gas storage is simple. And for cars that get 60 to 100 mpg, the volume and weight required for compressed gas storage at 2400 psia would be relatively small. To go 200 miles between refuellings, such cars could use hydrogen pressurized to 2400 psia and stored in three to five standard 50-liter compressed gas cylinders (containing 2 to 3.3 gallons of gasoline-equivalent energy).[134] This is a relatively modest volume, which could be accommodated in the trunk of even a small car. (Roughly 0.20 to 0.32 cubic meters would be required for 3 to 5 tanks, or 40 to 65 percent of the trunk space.) Today's natural gas-powered cars use this kind of storage system.

But filling station costs would be high for this storage option. It costs about $3 per GJ (or 37 cents per gallon of gasoline-equivalent) to compress a low-energy-density fuel like hydrogen from pipeline pressure (150 to 300 psi) to filling station storage pressure (3600 psi).[135] At higher storage pressures, filling station costs would be even higher.

Metal hydride storage (See Box 5) is perhaps the most widely researched and promising option for automobiles, one that has been pursued by Daimler-Benz among others. (See Figure 12.) In hydride storage, hydrogen is bound in the metal unless high temperature heat is applied to liberate it. Hydrides weigh about as much as compressed gas cylinders for the same amount of energy storage but take up only about one third as much space.[136] Much of the research on hydrides has centered on finding materials which can store a higher proportion of hydrogen on a weight basis. The difficulty of finding a hydride that can operate under the right range of conditions and still contains

more than 1 to 2 percent hydrogen by weight has led some researchers to abandon hydrides for liquid hydrogen. But, as with compressed gas, the problem of low energy density can be ameliorated by making the car more energy efficient. For a 100 mpg car, hydride storage for a range of 320 km would weigh 128 kg and take up 47 liters of space (about as much as a 12-gallon gasoline tank). Moreover, hydride tanks can be shaped like traditional gasoline tanks, and they are easier to fit into a car than are pressurized cylinders. The initial cost of hydride storage tanks would be only one third to one half that for gaseous storage, and costly high-pressure equipment would not be needed at filling stations.

Liquid organic hydrides are another possibility for onboard storage of hydrogen. This option is currently being investigated for large vehicles.[137]

Engines for Hydrogen-Powered Cars

Conventional spark-ignited (S-I), internal-combustion engines have been run on hydrogen in numerous prototype vehicles.[138] Hydrogen is not a suitable fuel for diesel engines, since it does not ignite easily under compression, but there are other automotive engines now under development that would be as efficient as diesels and are well suited to run on hydrogen fuel.[139] The most attractive of these appears to be the spark-ignited, direct-injection (DI) engine, sometimes called a stratified-charge engine. Direct-injection engines can be used with almost any fuel, and their energy efficiency is almost independent of fuel characteristics.[140] On an energy per kilometer basis, D-I engines operated on hydrogen, gasoline, natural gas or methanol (like Diesel engines run on Diesel fuel) would be about 50-percent more energy-efficient than current spark-ignited, gasoline-powered engines.

Hydrogen Fuel Distribution System

The fuel-distribution system for metal hydride-fueled cars could involve service sta-

Box 5. Metal Hydride Storage for Automobiles

Onboard storage of hydrogen in metal hydrides is a promising option for hydrogen-powered automobiles. Hydrogen reacts with many metals (among them iron, titanium, nickel) to form hydrides according to:

$$H_2 + Metal \rightarrow Metal\ Hydride + Heat$$

The reaction is reversible, if heat is added to the hydride:

$$Metal\ Hydride + Heat \rightarrow H_2$$

The temperature of heat required to liberate hydrogen depends upon the particular hydride. Metal hydrides fall into two groups "high-temperature hydrides" (HTHs), where heat at 300 to 400°C is needed, and "low-temperature hydrides" (LTHs), where the reactions begins when heat at 25 to 200°C is applied. As pressure is increased, a higher temperature is required for the reaction.

The HTHs can store about 7 to 8 percent of usable hydrogen by weight. The LTHs store about 1 to 2 percent hydrogen by weight. Because the HTHs would have a higher energy density than the LTHs, they would appear to be preferable. However, for automotive applications, it is desirable to produce hydrogen at pressures in the range of 10 to 50 atmospheres for injection into the engine, and to use the heat from the engine exhaust, which is generally less than 300°C during typical driving conditions, to liberate the hydrogen. Given these two considerations, the LTHs are more practical to use, even though their energy density is less. With HTHs, extra equipment for providing auxiliary heat would be needed, which would negate their advantage.

The storage tank for the hydride typically adds 15 to 25 percent to the weight of the hydride itself. Energy densities for the combined system (hydride plus tank) are 1.80 to 1.98 MJ per kg. Hydrides do not degrade appreciably with repeated charging, as long as the hydrogen is sufficiently pure. After 4000 cycles, the capacity of the material to combine with hydrogen is virtually unchanged. For a car with a 400 km range, 4000 charging cycles would be equivalent to 160,000 kilometers, about the lifetime of the car.

Source: H. Buchner, "Hydrogen Use—Transportation Fuel," *International Journal of Hydrogen Energy*, v. 9, pp. 501–514, 1984.

tions connected to a gas pipeline, the current practice for natural gas-powered cars. Daimler-Benz researchers estimate that it would take about 10 minutes per hydride storage tank "fill-up" at a delivery pressure of 735 psia. Filling station costs would add perhaps $1.5 per GJ (19 cents per gallon gasoline-equivalent) to the fuel cost, an amount comparable to filling station costs for compressed natural gas.[141] *(See Table 15.)*

The Economics of Cars Fuelled with Hydrogen

How much would it cost to own and operate a car fueled with PV hydrogen compared to the cost for other long-term options such as methanol, electricity, synthetic gasoline, synthetic natural gas, and batteries?

Alternative Fuel Costs. Table 15 shows the estimated production costs and delivered costs for

Table 15. Delivered Cost of Automotive Fuels circa 2000

	Long Distance Distrib.	Local Distrib.	Filling Station ($/GJ)	Total Delivery Cost	Production Cost[a]	($/GJ)	Cost to User ($/gallon gasoline equiv.)
Gasoline from coal	0.54	0.15	0.81	1.50[b]	16.15	17.65	2.30
Methanol from coal	0.79	0.30	0.93	2.02[b]	7.89[c]	9.91	1.29
					10.05[d]	12.07	1.57
Ethanol							
from corn, U.S.				1.49[e]	14.41	15.90	2.07
from sugar cane, Brazil					8.05	9.54	1.24
Synthetic natural gas							
from coal	1.10[f]	0.42[f]	1.57[g]	3.09	5.74[h]	8.83	1.15
					7.92[i]	11.00	1.43
PV Hydrogen		0.42[k]	1.45[g]				
$n=18\%$, \$0.2/W_p	1.92[j]			3.79	9.10	12.86	1.68
$n=12\%$, \$0.4/W_p	2.14[j]			4.01	13.96	17.97	2.35

a. *See Table 9, Notes 114 and 94.*
b. *Source:* "Preliminary Perspective on Pure Methanol for Transportation," EPA 460/3-83-003, 1982.
c. For a methanol synfuel plant producing 322,000 GJ/day. *See Note 114.*
d. For a methanol synfuel plant producing 32,200 GJ/day. *See Note 114.*
e. Assumed to be the same per liter as for methanol.
f. The costs of long distance transmission and local distribution are average values projected for all customers and industrial customers respectively for natural gas in the U.S. in 2000. *Source:* American Gas Association, Policy Evaluation and Analysis Group, Bulletin AGA-TERA 86-1, January 1986.
g. *See Note 135.*
h. For an SNG synfuel plant producing 264,000 GJ/day. *See Note 114.*
i. For an SNG synfuel plant producing 26,400 GJ/day. *See Note 114.*
j. Long distance transmission costs for hydrogen were estimated assuming transmission 1000 miles via pipeline. *See Note 100.*
k. As hydrogen and natural gas local distribution systems would be similar in size and handle about the same energy flow, we have assumed that hydrogen local distribution costs are the same as for natural gas.

various transportation fuels. If the cost and performance goals for PV modules set forth in Table 7 are met, PV hydrogen could be delivered to consumers for $12.9 to $18.0 per GJ ($1.68 to $2.35 per gallon of gasoline equivalent), making hydrogen somewhat more costly than methanol or synthetic natural gas (SNG) derived from coal, but probably less costly than gasoline derived from coal and ethanol derived from corn. (Hydrogen from PV sources would be much less expensive than electrolytic hydrogen from nuclear power, which would have a delivered cost of perhaps $18 to $28 per GJ or $2.35 to $3.65 per gallon of gasoline-equivalent.)

Consumer Costs for Cars with Alternative Fuels. In comparing the costs of owning and operating cars with alternative fuels, costs other than fuel should be taken into account: these include the capital cost of the vehicle (including any costs associated with modifying the vehicle to improve fuel economy and any costs for engine modifications or onboard storage tanks needed to accommodate alternative fuels), and maintenance costs, tolls, insurance, and parking costs.[142]

Figure 14 shows the total cost per kilometer of owning and operating a car powered by gasoline (@ $1 a gallon), methanol from coal, SNG from coal, PV hydrogen, and an electric battery. The comparison covers a range of fuel economies (corresponding approximately to 30, 50 and 90 mpg of gasoline-equivalent) and the range of delivered fuel prices presented in Table 15. As this figure shows, the non-fuel costs dominate total costs, even at 30 mpg. Fuel choice has little effect on the total cost to the consumer of owning and operating a car.

The National Cost of Alternative Transport Fuels. Although the consumer costs of owning and operating a car don't vary much for various synthetic transport fuels, the aggregate cost to the nation would be quite different for each option.

Table 16 illustrates how automobile transport fuel costs in the United States might evolve as vehicles and fuels change. (In this analysis, "automobiles" include light trucks as well as cars.) Specifically, this table shows the estimated annual automotive transport fuel cost circa 2020 with various synthetic fuel options for three levels of fuel economy, in relation to the actual expenditures on gasoline in 1984 and the range of expenditures projected for the year 2000 by the U.S. Department of Energy. Clearly, Table 16 shows that increasing automotive fuel economy would lead to dramatic reductions in the annual transport fuel

bill of the United States, even with many more vehicles on the road and more costly fuels.

What about Trucks, Buses, Trains, Ships, and Planes?

Buses, trucks, airplanes, and ships are also large oil users. In the United States in 1983, these energy "end uses" collectively accounted for 9.3 Exajoules (EJ) of oil use, about as much oil as automobiles used. *(See Table 13.)*

With these other transport modes, current fuel economy probably does not constrain the use of hydrogen, as it does with the automobile. With larger vehicles the weight and volume penalties associated with on-board hydrogen storage are not serious,[143] as has been demonstrated for hydrogen-fueled buses.[144] *(See Figure 12.)*

The total cost of owning and operating vehicles becomes less sensitive to fuel choice at higher levels of fuel economy, which makes it easier to select a fuel on the basis of external social costs as well as private costs.

As oil prices rise and shifts are made to more costly synthetic fuels, however, fuel economy improvements of these vehicles will become important for economic reasons. Fortunately, there are significant opportunities for making such improvements.[145] As in the case of the automobile, the total cost of owning and operating these vehicles becomes less sensitive to fuel choice at higher levels of fuel economy, which again makes it easier to select a fuel on the basis of external social costs as well as private costs.

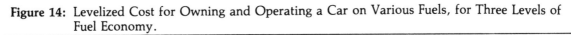

Figure 14: Levelized Cost for Owning and Operating a Car on Various Fuels, for Three Levels of Fuel Economy.

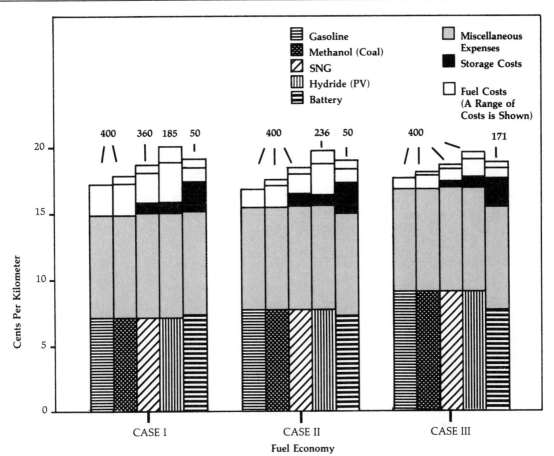

The levelized cost in cents per kilometer is shown for automobiles fueled with gasoline, methanol from coal, synthetic natural gas from coal, PV hydrogen, and electricity stored in batteries. The travelling range (in km) is shown at the top of the bar for each option.

Case I is roughly equivalent to a present day sub-compact car with a fuel economy of 30 miles per gallon of gasoline equivalent. Case II is roughly equivalent to a subcompact car with a more efficient engine (Diesel or stratified charge), with a fuel economy of 50 mpg gasoline. Case III corresponds to a 90 mpg car with a more efficient engine, aerodynamic styling and a continuously variable transmission. (A detailed description of each case is presented in Note 142.)

For each case, the levelized cost has four components: initial capital cost of the vehicle (which is shown in hatched patterns and includes the purchase price of the car, exclusive of extra storage system costs), miscellaneous expenses (shown in grey area above the initial capital cost of the vehicle, this includes tolls, registration fees, insurance, parking, repairs and maintenance), the storage cost (shown in black, this includes any extra cost for a special fuel storage system such as batteries, compressed gas cylinders or hydride tanks), and the fuel cost (shown in white at the tops of the bars). The delivered cost of fuels (from Table 15) are: $7.67/GJ ($1/gallon) for gasoline, $9.9-$12.1/GJ for methanol from coal, $8.8-$11.0/GJ for synthetic natural gas from coal, $12.9-$18.0/GJ for PV hydrogen, and $0.06-$0.10/kwh for electricity.

Table 16. The Automobile Fuel Bill of the United States

	Consumer Fuel Price ($/GJ)	Cost of Automobile Transport Fuel (Billion Dollars per Year)		
Gasoline from crude				
1984[a]	10.17	125		
2000 I[b]	7.97	103		
2000 II[b]	9.65	118		
2000 III[b]	11.93	137		
		Alternative Scenarios for 2020[c, d]		
		"30 mpg"	"50 mpg"	"90 mpg"
Gasoline from coal	17.65	153	92	51
Methanol from coal	9.91–12.07	73–89	55–67	30–36
SNG from coal	8.83–11.0	71–89	48–60	26–32
PV Hydrogen				
$n = 18\%$, $0.2/W_p$	12.86	95	74	41
$n = 12\%$, $0.4/W_p$	17.97	129	100	56

a. For a total light vehicle energy use of 12.3 EJ per year. Source: *Transportation Energy Data Book: Edition 9*, Oak Ridge National Laboratory, ORNL-6325, April 1987.

b. Alternative USDOE projections for gasoline consumption and prices in the year 2000, with alternative crude oil prices of $26.76/barrel for I, $32.87 for II, and $41.48 for III. *Source:* Energy Information Administration, U.S. Department of Energy, *Annual Energy Outlook with Projections to 2000*, DOE/EIA-0383(86), 1987.

c. We assume that there will be 235 million adults (aged 16 and over) in the U.S. in 2020, that there will be 0.80 light vehicles per adult, and that each vehicle travels 10,600 miles per year.

d. Based on fuel costs from Table 15, for three levels of average automotive fleet fuel economy corresponding roughly to 30, 50, and 90 miles per gallon of gasoline equivalent.

While there should be no formidable problems limiting the use of hydrogen in buses, trucks, trains, and ships, the use of hydrogen as an air transport fuel may be more problematic. Because hydrogen is readily burned in jet engines and has almost three times as much energy per unit weight as jet fuel *(See Table 14)*, hydrogen might seem attractive for large, long-range aircraft, where the fuel weight is a significant load. But for aircraft use, hydrogen would probably have to be liquefied to save space, and the expense of liquefaction (adding $6 to $10 per GJ to the cost of fuel) would make it more difficult for liquid hydrogen to compete with other synthetic fuels, at least in today's jumbo jets.[146]

Environmental Problems and Alternative Synthetic Fuels

Since all fuel-driven cars perform comparably at high fuel economy levels, and since the costs of owning and operating a car is about the same regardless of fuel, the choice of a transportation fuel for the long term need not be driven by private economic cost considerations alone. Instead, external social costs—such as the health risks associated with urban air pollution, the environmental risks due to acid deposition, and the risk of global climate changes due to carbon dioxide emissions—should enter heavily into the decision.

Substantial local and regional air quality improvements could be achieved by switching to hydrogen as a transport fuel. Moreover, if hydrogen were produced from PV power, no carbon dioxide would be emitted in fuel production. In contrast, as Table 17 and Figure 15 show, carbon dioxide emissions would increase sharply, if coal-based synthetic fuels were substituted for gasoline, unless there were a compensating improvement in automotive fuel economy.[147]

But a shift to PV hydrogen as an automotive transport fuel will not take place unless public policies are formulated to make the automotive fleet much more fuel-efficient and to reflect the view that PV hydrogen is preferable to the fossil fuel alternatives on environmental grounds. *(See Chapter 9.)*

The needed public sector interventions would cost the consumer little. Even to the nation, the economic penalty associated with choosing hydrogen instead of, say, coal-derived methanol, would be small and would decline sharply with increasing fuel economy. For example, with methanol costing $12.1 per GJ and hydrogen $15.4 per GJ, the penalty on the national fuel bill in 2020 would decline from $21 billion at 50 mpg to $14 billion at 90 mpg, and the total U.S. fuel bill for hydrogen for automobiles would still be only 40 percent of that for gasoline in 1984. *(See Table 16.)* On balance these interventions would add up to a fairly painless strategy for coping with a very serious problem.

Table 17. CO_2 Emissions for Alternative Automotive Fuels[a]

	CO_2 Emission Rate (kg/km)		
	25 mpg[b]	50 mpg[b]	100 mpg[c]
Crude oil → Gasoline + Diesel	0.251	0.125	0.063
Natural gas	0.167	0.084	0.042
Coal → SNG (Lurgi Dry Ash process)	0.394	0.197	0.099
Coal → Methanol (w/Texaco gasifier)	0.460	0.230	0.115
Coal → Gasoline, SNG, LPG (SRC-II process)	0.430	0.215	0.108
PV Hydrogen	0	0	0
PV Electricity → Battery	0	0	0

a. *See Note 147.*
b. Gasoline-equivalent fuel economy.

Figure 15: Automotive CO_2 Emissions with Various Fuels as a Function of Automotive Fuel Economy

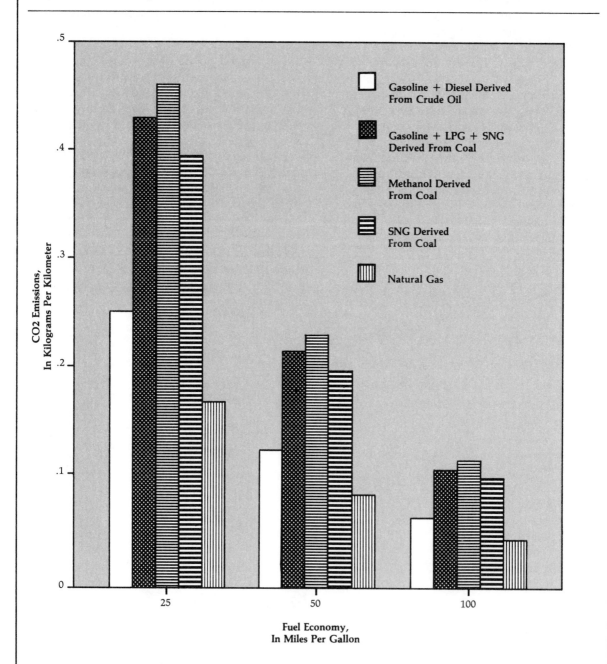

The number of kilograms of CO_2 released in the synthesis and combustion of fuel per kilometer travelled is shown for gasoline from crude oil, gasoline from coal, methanol from coal, and synthetic natural gas from coal. PV hydrogen would release no CO_2 to the atmosphere.

VII. Breaking into Markets for Gaseous Fuels

The need to begin a transition from natural gas to alternative fuels is probably less urgent than for oil. Since natural gas is the cleanest-burning fossil fuel, emphasizing it as an alternative to oil and coal can help alleviate local and regional air pollution problems. Moreover, increased use of natural gas is widely viewed as an important transition strategy for coping with the greenhouse problem, since carbon dioxide emissions are less for natural gas than for alternative fossil fuels. *(See Figure 3.)*

There is, however, another greenhouse gas problem that might constrain natural gas use in the decades immediately ahead—the problem of methane leakage. Methane is perhaps 25 times as effective as carbon dioxide as a greenhouse gas, so that leaks of methane gas to the atmosphere from natural gas production, transmission, distribution, and end-use systems in excess of a couple of percent of the gas produced would offset the greenhouse benefit associated with reducing carbon dioxide emissions by burning natural gas. The magnitude of such leaks at present is not well-known; neither is the extent to which leaks can be kept to acceptable levels, nor the effective lifetime of methane in the atmosphere.[148] But if methane leakage proves to be a serious problem, it could hasten the transition from natural gas to a more environmentally benign alternative fuel such as PV hydrogen.

In this context, what are the prospects for PV hydrogen in gaseous fuel markets, as an alternative to the production and use of synthetic gases from coal? Since synthetic natural gas (SNG) derived from coal is estimated to cost only $6 to $8 per GJ, compared to $9 to $14 per GJ for PV hydrogen *(See Table 9)*, PV hydrogen would seem less competitive in synthetic gaseous fuels markets than in synthetic liquid fuels markets. However, in some gaseous fuels applications, hydrogen will compete better than this production cost comparison would suggest, partly because the production costs of coal-derived gases depend strongly on the plant size. *(See Table 9.)* PV hydrogen may be economically competitive in applications where building large synfuels facilities is not feasible or practical. In addition, some of hydrogen's attractive properties may enable it to compete in some important applications even when its price on an energy-equivalent basis is considerably higher than that of coal-derived gases.

PV Hydrogen for Residential Heating

Residential space and water heating today account for about seven eighths of residential natural gas use. *(See Table 18)*. For these applications hydrogen can be compared to synthetic natural gas derived from coal and to electricity—the principal alternative energy carriers that might eventually replace natural gas.

Although hydrogen sounds like an exotic choice to most people in the United States

Table 18. World Natural Gas Consumption, in Exajoules/year (1983)[a]		
Market Economies		
United States		
Residential		5.0
Space Heat	3.29	
Water Heat	1.09	
Cooking, other	0.58	
Commercial		2.2
Space Heat	1.44	
Water Heat	0.36	
Cooking	0.16	
Cooling	0.06	
Other	0.13	
Industrial		7.1
Heat and Power	5.87	
Refinery Fuel	0.62	
Chemical Feedstock	0.52	
Electricity generation		3.6
Transportation		0.5
Total		18.6
Other OECD[b]		11.3
Developing Countries[c]		5.8
Centrally Planned Economies		
USSR		17.3
China		0.4
Others		3.2
World		56.5

a. *Sources:* British Petroleum Gas Company, *BP Preview of World Gas,* August 1985; American Gas Association, *The Gas Energy Demand Outlook 1984–2000,* 1984; American Gas Association, *Gas Facts,* 1986; U.S. Energy Information Administration, *Annual Energy Outlook, 1984,* DOE/EIA 0383(84), January 1985.
b. Other OECD includes Western Europe, Japan, Australia, New Zealand, Canada.
c. Africa, Southeast Asia, Pakistan, Latin America.

today, hydrogen-rich gases have been used for home heating and cooking for over a hundred years. "Town gas" (a mixture of approximately half hydrogen and half carbon monoxide, which can be derived from coal, wastes or wood) was piped into millions of urban homes in the United States earlier in this century, and it is still used in parts of Europe, Asia, and South America. In some U.S. regions, natural gas did not supplant town gas as a residential fuel until after World War II.

Although hydrogen sounds like an exotic choice to most people in the United States today, hydrogen-rich gases have been used for home heating and cooking for over a hundred years.

Hydrogen flame burners would resemble those in today's natural gas heating systems and appliances. The only important difference would be in the size of the burner openings controlling the velocity of the gas flow: hydrogen gas would have to flow three times as fast as natural gas to deliver the same amount of energy. Appliances using an open hydrogen flame would have to be vented to dissipate NO_x, as is done with natural gas flames. In fact, today's natural gas appliances could perhaps be converted to use hydrogen if the burners and some metering devices were replaced.[149] The efficiency, cost and performance should be essentially the same as with natural gas appliances.

Catalytic combustion offers the prospect of raising the efficiency of gas heaters. In a catalytic heater, fuel gas (hydrogen or natural gas) combines with oxygen in air at a relatively low temperature, in the presence of a catalyst such as platinum or stainless steel.[150] Instead of a flame, the catalytic reaction produces a radiant glow. Like electric space heaters, catalytic space heaters are installed in the room to be warmed, to take advantage of radiant heating.

Catalytic space heaters fueled with natural gas are commercially available for residential use.[151] These units are commonly used to heat a single room. They must be vented to the outside to avoid the buildup of such combustion products as carbon monoxide and NO_x, but they obtain heating efficiencies of 80 to 90 percent. With hydrogen fuel, catalytic combustion can be carried out at a temperature low enough to keep NO_x production negligible,[152] and the combustion products (mainly water vapor) can be discharged directly into the heated space. As a result, hydrogen catalytic burners can be close to 100 percent energy efficient! Moreover, the catalytic hydrogen heater can act as a humidifier, too, improving comfort by bringing the relative humidity up to 40 to 50 percent on cold days.[153] Alternatively, the steam from combustion could be condensed to make potable (distilled) water, rather than released into the room.[154]

Researchers in the United States, Japan and Europe have already constructed prototype catalytic space heaters and hot water heaters fueled with hydrogen.[155] The costs of hydrogen space heaters should be comparable to those of natural gas catalytic heaters today, and installation costs should be lower, as venting would not be neccessary.[156] Catalytic hydrogen water heaters would probably cost about 8 to 20 percent more than today's natural gas-fired storage tank-type heaters, but they would be 50 to 100 percent more energy-efficient![157]

How much would consumers pay for residential space and water heating with PV hydrogen compared to using synthetic natural gas (SNG) derived from coal or electricity? The answer to that question depends not only on residential fuel prices (See Table 19) but also on the energy end-use technologies involved (the type of space and water heaters and the design of the house).[158] In general, the economics of hydrogen improves markedly in relation to the alternative energy carriers as the efficiency of end-use technologies improves.

In Figure 16, the levelized annual cost of space and water heating with each fuel is sketched for three different levels of end-use energy efficiency.[159,160] Case I is a conventional house with low first-cost, but relatively inefficient space and water heating systems. Case II is a conventional house with energy-efficient heating systems. In Case III, the house's energy demand is lowered by using a "super-insulated" design, so that a much smaller heating system is required. (Super-insulated houses have extra insulation, double-glazed windows with "heat-mirror" coatings and other features which cut the annual space heating demand to about one eighth that of a comparable conventional new house. A super-insulated house costs about $4100 more than a comparable quality conventional house.[161]) The costs and performance for each case are given in Table 20.

How do the various options compare in cost? When low first-cost end-use technologies are used in conventional houses, SNG systems are the least costly, followed by electricity, with hydrogen coming in last. When energy-efficient space and water heaters are installed in conventional houses, overall costs are reduced for hydrogen and electric systems, though not for SNG systems, and hydrogen fares somewhat better relative to the other energy carriers. If hydrogen at the low end of its estimated cost range were used, hydrogen-fueled systems could provide heat at about the same cost as for electrical systems and SNG systems fueled with SNG priced at the high end of its estimated cost range. When energy-efficient end-use technologies are used in superinsulated houses, overall costs drop sharply for all fuels—even though superinsulated houses are more costly to build. Surprisingly, hydrogen becomes competitive in this instance. In a super-insulated house, using hydrogen is clearly less expensive than using electricity for space and water heating. Even with hydrogen priced at the high end of its estimated cost

Table 19. Delivered Cost of Residential Heating Fuels ($/GJ)

Fuel	Production Cost[a]	Long Distance Distribution	Local Distribution	Delivered Cost to Consumer
Synthetic natural gas from coal	5.79→7.92	1.12[b]	1.50[b]	8.41–10.54
PV Hydrogen			1.50[d]	
$n=18\%$, $0.2/W_p$	9.07	1.92[c]		12.49
$n=12\%$, $0.4/W_p$	13.96	2.14[c]		17.60
Electricity				16.67–27.78[e] ($0.06–$0.10/kWh)

a. See *Table 9* and *Note 114* for SNG from coal and *Note 94* for PV hydrogen.
b. American Gas Association, Policy Evaluation and Analysis Group, Bulletin A.G.A.-TERA 86-1, January 1986.
c. This estimate is for transmitting hydrogen 1000 miles, with an assumed pipeline inlet pressure of 1000 psia and pipeline outlet pressure of 300 psia. *See Note 100.*
d. We assume that the local distribution of natural gas and hydrogen would involve similar systems and that the costs would be the same.
e. The average residential electricity price projected for the year 2000 is $0.074/kWh. Energy Information Administration, U.S. Department of Energy, *Annual Energy Outlook with Projections to 2000*, DOE/EIA-0383(86), 1987.

range, annual space and water heating bills would be about 15 percent less costly than with electrical systems. Hydrogen would be competitive with SNG at the high end of the price range and about 5 percent less costly at the low end of the range.

This remarkable shift in the economics of hydrogen arises because of the capital savings and efficiency improvements that are possible in end-use equipment used with super-insulated housing designs. When hydrogen is used to heat the conventionally constructed house, a complex centralized heating system costing some $3750 installed is required. In contrast, when hydrogen is used to heat the super-insulated house it is feasible to meet heating needs with three small space heaters costing just $855. Of course, there is also a large reduction in the capital cost of heating equipment with the SNG system—but the required catalytic heaters are somewhat more costly ($1245 vs. $855) and less efficient (85 percent versus 99 percent), because the SNG units must be vented.

PV Hydrogen for Industry

Hydrogen for Industrial Heat and Power

Hydrogen would probably face tougher competition in the industrial process heat and power-generation gaseous fuels markets, which account for about half of U.S. natural gas consumption, than in residential markets. *(See Table 18.)* Here the synthetic gaseous fuel of choice may well be gas of intermediate heating

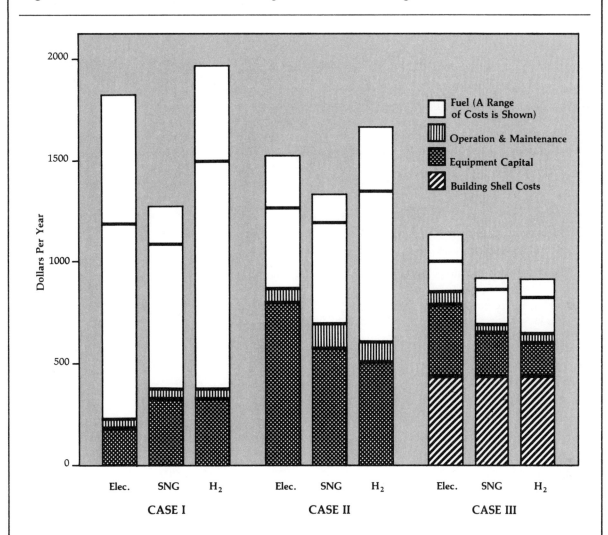

Figure 16: Levelized Costs for Residential Space and Water Heating.

The costs are for houses located in New Jersey and heated with electricity, synthetic natural gas from coal, and PV hydrogen for three levels of energy end-use technology.

Case I involves low first-cost equipment in conventionally-constructed houses. Case II involves the use of energy-efficient equipment in conventionally-constructed houses. Case III involves the use of energy-efficient equipment in super-insulated houses.

The levelized cost has four components: building shell tightening costs (which apply to the super-insulated house), heating system capital costs, operation and maintenance costs and fuel costs.

A range of fuel costs are shown, corresponding to high and low energy prices from Table 19. The heat requirements for space and water heating are estimated in Note 158. The assumed capital and O&M costs are presented in Table 20. A ten percent discount rate is assumed for the lifecycle cost calculations. See Note 159.

Table 20. Cost and Performance of Natural Gas, Hydrogen and Electric Heating Systems for Conventional and Super-Insulated Houses in New Jersey[a]

	Size	Costs ($) Installed Unit	Ducts	Vents	System Total	Effic. or COP	Life (yr)	O&M ($/yr)
1. Low First-Cost Technologies, Conventional New Houses								
NATURAL GAS:								
Conventional furnace	41 kBTU/h	700	1500	50	2250	0.69	18	25
Tank-type water heater	40 gal	320	–	50	370	0.52	10	5
HYDROGEN:								
Conventional furnace	43 kBTU/h	700	1500	50	2250	0.65	18	25
Tank-type water heater	40 gal	320	–	50	370	0.49	10	5
ELECTRICITY:								
Resistance space heat	8.28 kW	1100	–	–	1100	0.99	15	15
Resistance water heater	40 gal	300			300	0.85	10	5
2. Energy-Efficient Technologies, Conventional New Houses								
NATURAL GAS:								
Recuperative furnace	29 kBTU/h	2200	1500	50	3750	0.95	20	80
Tankless water heater	40 gal	750	–	50	800	0.78	10	30
HYDROGEN:								
Recuperative furnace	29 kBTU/h	2200	1500	50	3750	0.95	20	80
Catalytic water heater	40 gal	425	–	50	475	0.83	10	5
ELECTRICITY:								
Heat pump	2.87 kW	2700	1500	–	4200	2.85	15	25
Heat pump water heater	40 gal	1550			1550	1.8	10	30
3. Energy-Efficient Technologies, Superinsulated Houses								
NATURAL GAS:								
Tightened building shell					4100			
3 Catalyt. sp. htrs	11.5 kBTU/h	1245	–	–	1245	0.85	15	10
Tank-type water heater	40 gal	320	–	50	370	0.52	10	5
HYDROGEN:								
Tightened building shell					4100			
3 Catalyt. sp. htrs	9.9 kBTU/h	855	–	–	855	0.99	15	10
Catalytic water heater	40 gal	425	–	50	475	0.83	10	5
ELECTRICITY:								
Tightened building shell					4100			
Resistance heaters	2.89 kW	800	–	–	800	0.99	15	15
Heat pump water heater	40 gal	1550	–	–	1550	1.8	10	30

a. The technical characteristics of the space and water heating systems are described in Note 159. The equipment capital and O&M costs are from the Electric Power Research Institute's *Technical Assessment Guide, Volume 2: Electricity End Use, Part I: Residential Electricity Use,* EPRI, P-4463-SR, September 1987, for all but the catalytic heaters; cost estimates for catalytic heaters are from B. Wells, Thermal Systems, Inc., Tumwater, Washington, private communication, 1987.

value (consisting mainly of carbon monoxide and hydrogen) derived from coal, which can probably be produced for about $4 per GJ, considerably less than the cost of synthetic natural gas or hydrogen. *(See Table 9.)*

As in the residential case, however, the relative production cost may not always be the best indicator of competitive advantage. U.S. industry consumes about two fifths as much electricity as natural gas, even though electricity is nearly five times as costly per GJ. Electricity is preferred and often necessary for many applications, despite its higher price, because it is a more versatile energy source. Likewise hydrogen's properties may make it preferable to coal gas in some applications.

Consider one example. Because PV electrolysis can be carried out economically on a small scale, large industrial users in sunny areas might choose to operate their own PV electrolysis units as an alternative to buying coal gas from the utility. Utility transmission and distribution costs (which could total perhaps $1.5 per GJ to industrial users of synthetic gas) would be avoided and a valuable byproduct, oxygen (which is worth perhaps $1.5 to $2.2 per GJ of hydrogen[162]) would be produced. (Of course, the relative economics of coal gas and PV hydrogen would also depend on the need for storage, the load factor of the process and a host of other factors.)

PV Hydrogen as a Chemical Feedstock

Hydrogen, a basic chemical building block, can be combined with a carbon source such as coal or biomass to form a myriad of useful compounds. In some chemical processing industries where natural gas (CH_4) is used primarily as a source of hydrogen, hydrogen produced via PV-powered electrolysis may prove economically attractive in some applications. *(See Box 7.)*

Summary

Perhaps the most promising potential gaseous fuel market for PV hydrogen is residential space and water heating—which accounted for about one eighth of U.S. energy use in 1980 (10 EJ). While the PV hydrogen prices projected here for the residential sector would seem to preclude the development of this market, the price of hydrogen is not a good indicator of this fuel's market penetration potential. A much better indicator is the *lifecycle cost of energy services that would be provided by hydrogen fuel.* The lifecycle cost comparison shows that hydrogen offers distinct advantages in terms of energy efficiency and capital equipment requirements that can be readily exploited if energy-efficient end-use technologies are adopted.

Perhaps the most promising potential gaseous fuel market for PV hydrogen is residential space and water heating— which accounted for about one eighth of U.S. energy use in 1980.

Hydrogen will probably be used in residential heating only after it has become well-established as a transportation fuel. The first PV hydrogen systems in the United States will probably be built to serve local transport markets in the Southwest, where space heating needs are minor but the need for clean automotive fuels is crucial. If hydrogen pipelines were built linking PV production facilities to transport markets in the Northeast and Midwest, residential heating markets might subsequently develop for hydrogen.

VIII. An Evolutionary Path to the PV Hydrogen Economy

Concerns about the local, regional and global environmental impacts of fossil fuels, as well as the prospect of declining domestic oil production, are prompting decision-makers to pay increasing attention to less-polluting, domestically-produced alternative fuels. The most notable actions to date along these lines in the United States are the passage of the Alternative Motor Fuels Act of 1988, which relaxes the automotive fuel economy standards for cars capable of using alternative fuels, and the California South Coast Air Quality Management District's 1988 proposal to phase out petroleum in Southern California in favor of alternative fuels, to help control air pollution. To the extent that they would promote a shift to methanol, these actions could help cope with oil import dependency and local air pollution problems, though they would exacerbate the greenhouse problem. Nevertheless, these actions underscore the gravity of the underlying issues in the minds of decision-makers.

Near the turn of the century, photovoltaic hydrogen could begin to compete economically in these emerging alternative fuels markets, offering an option that could help meet local and regional air pollution control goals and reduce oil import dependency without contributing to the global greenhouse warming. But how would a PV hydrogen energy system evolve in the United States?

PV Hydrogen for Chemical Markets

As has often been suggested, the first economically viable markets for PV hydrogen might be niches in present chemical hydrogen markets rather than in the transport fuels markets. Today, hydrogen is used primarily in the production of ammonia, methanol and various chemical products and in oil refining. *(See Box 6.)* Hydrogen can be produced from fossil fuels or via water electrolysis. *(See Box 7.)* Within broad limits, the smaller the daily hydrogen demand, the higher the cost of hydrogen to the user. *(See Figure 17.)* In the United States, at present, steam reforming of natural gas is the least expensive and most widely used method of large-scale hydrogen production. Small users generally buy "merchant" hydrogen from suppliers.

The first economically viable markets for PV hydrogen might be niches in present chemical hydrogen markets rather than in the transport fuels markets.

77

Today hydrogen is used primarily as a chemical feedstock for ammonia synthesis, oil refining, methanol production, and smaller industrial uses including metals reduction, hydrogenation of fats and oils, electronics, glass production, pharmaceuticals, and electric generator cooling. In some areas of the world, hydrogen-rich synthetic gases (from coal, wastes or biomass) are burned as fuel.

Hydrogen use worldwide is about 1.4 Exajoules per year, of which 62 percent is for industrial purposes other than energy (e.g. the production of ammonia, methanol, and other industrial products), 34 percent is for indirect energy use (petroleum refining), and 4 percent is direct energy use (as a component of synthetic fuel gas).[163] Figure B.6.1 shows hydrogen production in various countries as a percentage of total energy use.

The United States accounts for about half of global hydrogen consumption, some 0.65 EJ per year—equivalent to almost 1 percent of total U.S. energy use. Table B.6.1 lists the main uses of hydrogen in the U.S., the associated annual demand, the source of this hydrogen, and the approximate onsite production cost (See Box 7)[164] or price (for "merchant hydrogen" purchased from a supplier).[165]

Over the next ten to twenty years, opportunities for commercializing PV hydrogen for chemical use could be significant. (See Box 8.) For small-volume users in sunny areas, PV-powered electrolysis could begin to compete with merchant hydrogen or small steam-reforming plants in the early to mid–1990s. To supply the entire U.S. merchant hydrogen market would require perhaps 400 MW of PV capacity. For intermediate-sized users located in sunny regions, representing a total market of perhaps 3000 MW of PV capacity, PV hydrogen is likely to become competitive in the late 1990s. Large chemical markets would probably open around the same time as energy markets. To supply this entire market would require about 60,000 MW of PV capacity.

Introducing PV Hydrogen as a Transportation Fuel

While chemical markets could play a role in the early development of PV hydrogen, the major environmental benefits offered by PV hydrogen would begin to be realized only when PV hydrogen begins to penetrate transportation energy markets. When could the transition to a hydrogen-based transportation system begin?

Getting Started: a Hydrogen-Based Transport System for Phoenix?

A promising strategy for initiating any new technological system would be to identify an initial application where costs are favorable, and the need is pressing. The system should also be able to be introduced gradually, one small step at a time. To illustrate how this approach might be used to initiate a hydrogen economy, consider a hypothetical PV hydroen-based transport system for Phoenix, Arizona.

Since Phoenix is one of the sunniest areas in the United States, it is a good candidate for economical PV hydrogen production. And because the hydrogen would be used in Phoenix itself, the costs and complications of long-distance hydrogen transport could be deferred.

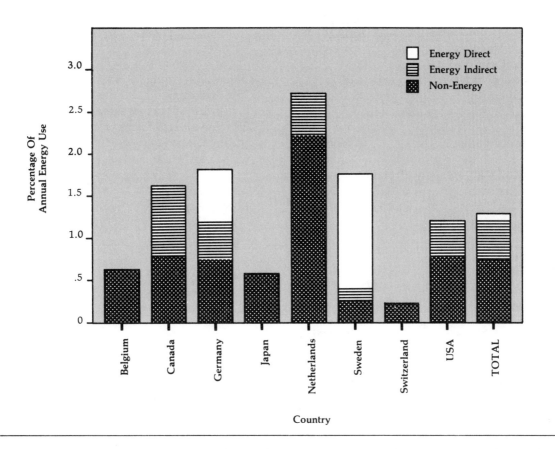

Figure B.6.1. Annual Hydrogen Production in Various Countries as a Percentage of Total Energy Use.

Source: W. Balthasar, "Hydrogen Production and Technology: Today, Tomorrow and Beyond," *International Journal of Hydrogen Energy*, v. 9, p. 649-668, 1984.

Since Phoenix is one of the sunniest areas in the United States, it is a good candidate for economical PV hydrogen production.

Clean transport fuels like PV hydrogen are greatly needed in Phoenix to cope with rapidly deteriorating urban air quality. *(See Box 9.)* In 1987, Phoenix violated federal air quality standards for ozone, carbon monoxide, and particulates on 33 days. Passenger vehicles were responsible for most of the ozone and particulates and for 80 percent of the carbon monoxide.[176]

Unless automobile emissions are controlled, air pollution in Phoenix could become much

Table B.6.1. Hydrogen Use in the United States

Use	Annual Hydrogen Demand (EJ/yr)[b]		Hydrogen Production Method[c]	Hydrogen[a] Cost ($/GJ)
LARGE USERS[d]				
Ammonia Synthesis	0.393		Steam Reforming	7.5–10.2
Oil Refining	0.167		75% Recovery[e] from off gas streams 25% Steam Reforming	
Methanol Synthesis	0.0666		Steam Reforming	
SMALL USERS[f]	Total	Merchant		
Misc. Chemical	0.0177	0.0011	Small Reformers	8.5–36
Metals Reduction	0.00360	0.00043	or Merchant	18–100[g]
Hydrogenation of fats and oils	0.00292	0.00026	''	
Electronics	0.00076	0.00076	''	
Float Glass	0.00032	0.00025	''	
Pharmaceutical	0.00022	0.00022	''	
Other Merchant	0.00029	0.00029	''	
Total Small Users	0.0258	0.00327		
TOTAL ALL USERS	0.653	0.00327		

a. Hydrogen production costs are estimated in Note 164.
b. The hydrogen energy demand is given on a higher heating value basis.
c. W. Balthasar, "Hydrogen Production and Technology: Today, Tomorrow and Beyond," *International Journal of Hydrogen Energy*, v. 9, pp. 649–668, 1984.
d. For 1975 hydrogen use in the U.S.. E. Fein and T. Munson, "An Assessment of Non-Fossil Hydrogen," Gas Research Institute Report, GRI 79/0108, December 1980.
e. C.R. Baker, "Production of Hydrogen for the Commercial Market: Current and Future Trends," American Chemical Society Report, 0-8412-0522-1/80/47-116-047-229, 1980.
f. For 1977 hydrogen use. R.W Foster, R.R. Tison, W.J.D. Escher, and J.A.Hanson, "Solar/Hydrogen Systems Assessment," DOE/JPL-955492, 1980.
g. For merchant hydrogen, the price rather than the production cost is given. E. Fein and K. Edwards, "Market Potential of Electrolytic Hydrogen Production in Three Northeastern Utilities' Service Territories," EPRI Report, EM-3561, May 1984.

worse. The population of the greater Phoenix area is projected to grow from two million people in 1985 to three million by the year 2000, and to five million by 2015. During this time, automobile use is projected to more than triple from 38 million vehicle-miles per day in 1985 to 119 million vehicle-miles per day in 2015.[177]

If the transition began with automotive fleet vehicles in Phoenix, the transition could occur

Box 7. Hydrogen Production Methods and Costs*

Steam reforming of natural gas. In this most widely used method of hydrogen production, a light hydrocarbon such as natural gas or naphtha reacts with steam in the presence of a nickel catalyst to produce hydrogen and carbon monoxide. For natural gas prices in the range $4 to $6 per GJ, the production cost of hydrogen by this method would be $7.5 to $10.2 per GJ. Almost all ammonia and methanol plants produce hydrogen by steam reforming.

Partial oxidation of residual oil. In this process, heavy oil is gasified with steam and oxygen. The synthesis gas (a mixture consisting mainly of carbon monoxide and hydrogen) produced this way is processed through several steps to give hydrogen and byproduct sulfur. For a residual oil price of $4 to $6 per GJ ($25 to $37 per barrel) the cost of hydrogen would be about $11.2 to $13.6 per GJ.

Recovery of hydrogen in oil refining. About 75 percent of the hydrogen required for oil refining is recovered from the end products of other chemical reactions. (The cost of hydrogen recovery depends on the particular plant design, but would often be less than for alternative methods of hydrogen produc-

tion.) The remaining 25 percent is produced by steam reforming of natural gas or naphtha or by partial oxidation of oil.[166]

Coal gasification. Rapid partial oxidation of pulverized coal can be used to form hydrogen. For a coal price of $1.78 per GJ (the average price projected for steam coal in the U.S. in the year 2000),[167] the cost of hydrogen would be $7.9 to $13.9 per GJ. Biomass gasification is another potential source of hydrogen.[168]

Water electrolysis. At present, only about 1 percent of the world's hydrogen is produced electrolytically.[169] Electrolysis is used in several large-scale plants where inexpensive hydroelectricity is available.[170] As shown in Chapter 4, if DC electricity is available at $0.020 to $0.035 per kWh, the cost of large-scale PV-powered electrolytic hydrogen production (exclusive of storage, compression or transmission) in the Southwest would be about $9.1 to $14.0 per GJ. For PV electrolysis there is little economy of scale above 5 to 10 MW (or 0.3 to 0.6 million standard cubic feet per day).

____

*Hydrogen production costs are estimated in Note 164.

---

in small steps. The initial demand for hydrogen could be satisfied using low-cost, off-peak power from conventional power plants well before PV hydrogen is commercially ready. This way, considerable field experience could be gained with hydrogen-powered vehicles before PV hydrogen is introduced on a much larger scale.

Table 21 shows how three alternative courses of action would affect automotive emissions in Phoenix. In Scenario I, emissions levels per car

remain fixed at the current average levels, and automotive pollution quickly becomes intolerable. By 2000, emissions from cars increase by about 50 percent, and by 2015, they triple, implying perhaps a doubling in the rate of ozone production by 2015. This scenario highlights the urgency of new measures for reducing automotive emissions.

Scenario II shows what would happen to emissions levels if recently proposed automotive emissions standards were implemented.

Figure 17: Hydrogen Production Costs Versus Daily Hydrogen Production Volume and Merchant Hydrogen Prices Versus the Daily Volume Purchased.

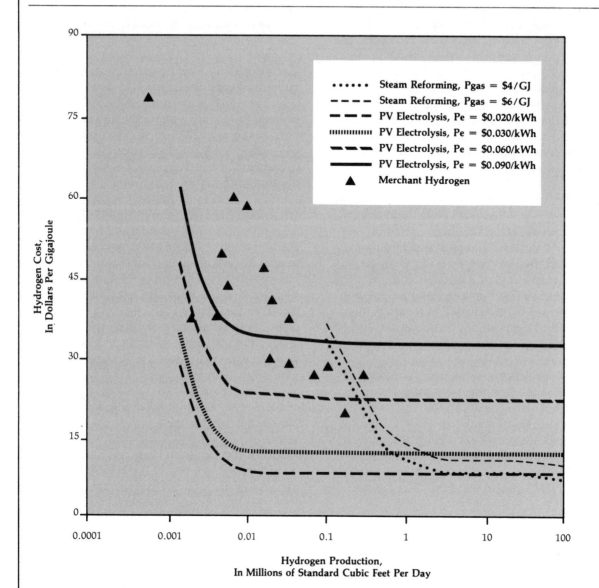

Costs for the steam reforming of natural gas are shown for natural gas prices of $4 to $6/GJ. Costs for PV electrolytic hydrogen are shown for Southwest U.S. conditions and DC electricity costs of $0.020 to $0.090/kWh. See Note 164.

The merchant hydrogen prices shown are based on an EPRI survey (see E. Fein and K. Edwards, "Market Potential of Electrolytic Hydrogen Production in Three Northeastern Utilities' Service Territories", EPRI Report, EM-3561, May 1984).

Box 8. PV Hydrogen for Chemical Markets

When could PV hydrogen begin to compete in chemical markets? The answer is implicit in Figure 17, where PV hydrogen production costs (in the Southwest) are compared to costs for hydrogen produced by steam reforming of natural gas and to merchant hydrogen prices for a range of hydrogen production capacities from 0.001 to 100 million standard cubic feet per day (scf per day). (One million scf per day = 339 GJ per day = 56 barrels per day of oil-equivalent.) The cost of PV-powered electrolysis is plotted against production capacity for several DC electricity costs ranging from $0.020 to $0.09 per kWh, corresponding to various PV module efficiencies and costs. The cost of steam reforming is plotted for natural gas prices of $4 to $6 per GJ. Both electrolysis and steam reforming exhibit economies of scale, though these are less severe for electrolysis, because of its modular nature. The prices plotted for merchant hydrogen are based on a recent EPRI survey of small-volume users of hydrogen.[171] Although merchant prices vary widely, in general, the smaller the volume of hydrogen use, the higher the merchant price paid.

The Merchant Hydrogen Market

Because merchant hydrogen customers pay premium prices ($18 to $100 per GJ) for hydrogen delivered by truck, it has been suggested that electrolytic hydrogen could compete in the merchant hydrogen market in the United States.[172] Merchant hydrogen users concerned about reliability of supply and hydrogen purity might have an added incentive to buy their own electrolyzers.[173]

For merchant hydrogen users with demands less than about 0.03 million scf per day (10 GJ per day), PV-powered electrolysis would be competitive for electricity costs of less than about $0.09 per kWhDC. PV electricity could be produced in the southwestern U.S. for $0.09 per kWhDC using 12-percent efficient modules with balance of systems (BOS) costs of $50/m² and PV module costs of $1.4 per peak Watt. *(See Figure 8.)* This system is within the range of projections for the early to mid–1990s. *(See Table 3.)* For those merchant hydrogen customers located in sunny areas now paying $30 to $80 per GJ, PV hydrogen could begin to compete within a very few years.

Because PV hydrogen technology is new, it may take an usually innovative small company to try this method as a replacement for merchant hydrogen. Large chemical, aerospace, or energy companies, automobile manufacturers or utilities with long-term interests in hydrogen research and development may be the first to exploit chemical market opportunities, as a way of helping finance PV hydrogen demonstration projects. This approach is being taken in Germany, where BMW, GmbH, and various utilities are now building an experimental 600 kW PV hydrogen system that will supply fuel for BMW's research on hydrogen-powered cars.[174]

Larger consumers of merchant hydrogen with demands of 0.03 to 0.3 million scf per day (10 to 100 GJ per day) pay $15 to $30 per GJ. PV-powered electrolysis could compete in these markets when electricity costs less than about $0.06 per kWhDC. A PV system with 12-percent efficient modules and a BOS cost of $50 per m² and a module cost of $0.8 per peak Watt could produce electricity at this cost. *(See Figure 8.)* By the mid–1990s, such systems may well be available.

The total U.S. merchant hydrogen market is about 0.003 EJ per year *(see Table B.6.1.)*.

While this is tiny compared to energy markets, about 400 MW of PV capacity would be required to supply all U.S. merchant hydrogen via PV-powered electrolysis. For comparison, in 1987, world production of PV systems was about 30 MW per year.

Intermediate-sized chemical markets

Medium-sized chemical users who make their own hydrogen using small natural gas reformers typically pay $12 to $20 per GJ. Figure 17 shows that to compete with small natural gas reformers for this market, with demand levels in the range 0.3 to 1.0 million scf per day (100 to 339 GJ per day), PV electricity must cost less than about $0.03 to $0.05 per kWhDC. This electricity cost corresponds to 12-percent efficient modules with BOS costs of $33 per m^2 and module costs of $0.3 to $0.6 per peak Watt—costs and efficiencies projected for a-Si solar cells sometime in the late 1990s.

Electrolysis would produce not only hydrogen but also high purity oxygen. Chemical users might be able to sell this oxygen as a byproduct, earning up to $1.5 to $2.2 per GJ of hydrogen produced. If oxygen could be sold, less stringent PV cost and efficiency goals would be needed to make PV hydrogen competitive with steam reforming.

The overall size of this hydrogen market is estimated to be some 0.026 EJ per year—almost 10 times as large as the merchant hydrogen market. Perhaps 3000 MW of PV capacity would be required to make this much hydrogen every year. Again, this could be a potentially important market for PV hydrogen systems in the mid- to late-1990s.

The large chemical markets

Large users (ammonia and methanol plants and oil refineries) produce their own hydrogen (primarily by steam reforming of natural gas) at a cost of $7.5 to $10.2 per GJ, for natural gas prices of $4 to $6 per GJ.[175]

By the turn of the century or shortly thereafter, PV hydrogen produced in the southwestern United States could cost $9.1 to $14.0 per GJ for 12- to 18-percent efficient PV modules costing $0.20 to $0.40 per peak Watt. If an oxygen byproduct credit of $1.5 per GJ were received, the net cost of PV hydrogen would be $7.6 to $12.5 per GJ, making it approximately competitive with reforming in this time frame.

(See Box 9 and Chapter 2). The proposed standards could be achieved by a combination of technical improvements and a rigorous maintenance-and-inspection program. In this scenario, automotive emissions fall to about half the present level by the year 2000. But even these measures would provide only temporary relief. By 2015, total automotive emissions return to the troublesome 1988 levels because of the growing number of cars. Scenario II highlights the importance of switching in the longer term from gasoline to low-polluting fuels such as hydrogen.

In Scenario III, tightened emission standards are combined with a gradual shift to hydrogen cars, so that by the year 2000 half of automotive fleets (60,000 cars, accounting for 4 percent of all automobiles but over 10 percent of all automobile miles) are operating on hydrogen, and by the year 2015 half of all cars in Phoenix are operating on hydrogen. In this scenario,

Box 9. Hydrogen Fuel and Automotive Emissions

In shifting from gasoline (see Table B.9.1) to hydrogen, carbon monoxide, particulate and VOCs emissions would be eliminated, and NO_x emissions could be reduced. NO_x emissions from laboratory hydrogen engines have been made much lower than those from comparable gasoline engines by using very lean fuel mixtures, exhaust gas recirculation, or water injection. (A few studies of actual on-the-road emissions from hydrogen vehicles during standard driving cycles show that: while measured emissions vary with vehicle size and installed pollution control equipment, NO_x emissions from hydrogen cars could be perhaps 10 to 20 percent of those from equivalent gasoline-powered

cars.) To date experimental hydrogen vehicles have been based on converted gasoline engines, and further reductions in NO_x formation may be possible with engines optimized for hydrogen. The assumption here is that NO_x emissions from compact (30 to 50 mpg) hydrogen-powered cars are 0.24 grams per mile. This value is based on laboratory engine tests and is consistent with measured actual emissions from road tests. (See M.A. DeLuchi, "Hydrogen Vehicles," *International Journal of Hydrogen Energy*, v. 14, No.2, 1989, pp. 81–130.) For a hydrogen-powered car using a fuel cell coupled to an electric motor, there would be no significant NO_x production.

Table B.9.1 Emissions from Gasoline-Powered Cars (in grams/mile)[a]

	AVERAGE EMISSIONS		EMISSIONS STANDARDS	
	Certified (1987)	Actual (1986)	1988	Proposed
Automobiles (below 6000 lb.)				
Carbon Monoxide	1.91	10.0	3.4	3.4
Hydrocarbons (VOCs)	0.20	0.8	0.41	0.25
Nitrogen Oxides	0.37	1.2	1.0	0.4
Particulates	–	–	0.2	0.08

a. *Source:* Michael P. Walsh, "Pollution on Wheels," Report to the American Lung Association, February 11, 1988.

automotive emissions fall to about half the present levels by 2000 and are still well below present levels by the year 2015.

When could conversion to hydrogen fuel begin? In the early to mid 1990s, PV hydrogen production (excluding storage and filling station costs) is projected to cost $18 to $32 per GJ, (equivalent to gasoline costing $2.4 to $4.1 per gallon), which may be too costly to justify

a large-scale fleet conversion program. PV hydrogen would not become cost competitive with other alternative fuels such as methanol until about the year 2000. (See Table 7.) Clearly, a less costly source of hydrogen may be needed to launch a hydrogen car program before the turn of the century.

As long as the initial vehicle fleet is not too large, enough low-cost, off-peak power would

Table 21. Alternative Scenarios for Automotive Emissions in Phoenix

	1985	2000	2015
Demographic Trends for Phoenix			
Population (millions)	2	3	5
Number of automobiles (millions)	1	1.5	2.5
Automobile-miles per day (millions)	38	57	119
Total Automotive Emissions (tonnes/day)			
SCENARIO I: No Change in Emissions per Car[a]			
CO	380	570	1190
Hydrocarbons	30.4	45.6	95.2
NO_x	45.6	68.4	142.8
SCENARIO II: Proposed Emissions Standards[b]			
CO	380	168	404
Hydrocarbons	30.4	14.3	29.8
NO_x	45.6	22.8	47.6
Particulates	–	4.6	9.5
SCENARIO III: Proposed Emissions Standards + Gradual Shift to Hydrogen[c]			
CO	380	147	202
Hydrocarbons	30.4	12.8	14.9
NO_x	45.6	21.6	38.1
Particulates	–	4.2	4.8

a. Based on 1986 in-use emissions. *(See Table B.9.1.)*
b. Assuming the adoption and enforcement of the proposed standards *(see Table B.9.1.)* for the years 2000 and 2015.
c. Assuming that the proposed standards are adopted and enforced for gasoline cars, that 50% of automotive fleets (4% of all cars but 10.5% of all vehicle miles) are operating on hydrogen by 2000, and that 50% of all cars are operating on hydrogen by 2015.

be available from conventional existing power plants in Phoenix to serve as an interim source of power for electrolytic hydrogen production. As much as several hundred megawatts of power could be available at $0.02 to $0.03 per kWh between the hours of 10 p.m. and 7 a.m.[178] With power at this cost, electrolytic hydrogen could be produced and delivered to the consumer for $12 to $16 per GJ (equivalent to $1.6 to $2.1 per gallon of gasoline),[179] which is approximately the cost targeted for PV hydrogen delivered to consumers in a full-blown PV-based hydrogen economy for automobiles. *(See Table 15.)* This cost estimate neglects any credit for byproduct oxygen, which could reduce the cost of the hydrogen by $1.5 to $2 per GJ (a credit equivalent to saving $0.20 to $0.29 per gallon of gasoline).

If 480 MW of off-peak power were available for hydrogen production, it would be possible to produce enough fuel for 30,000 fleet vehicles, each travelling 100 miles per day, thus accounting for about 5 percent of automobile miles in Phoenix in 2000,* assuming an average gasoline-equivalent fuel economy of 30 mpg. At this fuel economy, a typical metal hydride car would have a range of 115 miles,[181] so that the fleet cars could be operated during the day and evening and refuelled at night. The total capital cost for this initial system would be about $132 million for the electrolyzers, hydrogen compressors, gaseous hydrogen storage, and associated refuelling stations.[182] This amounts to about $1260 per fleet vehicle served by this system.[183] The metal hydride storage system would probably cost an additional $370 per fleet vehicle.[184]

To put the investment required into perspective, a recent study suggested building an extensive public transportation system for Phoenix at a cost of over $7 billion dollars to cope with the air pollution and congestion challenges associated with the expected population explosion.[185] While the new system would doubtless ease congestion and improve the quality of life for its users, it would reduce the number of automobile-miles travelled by only about 2 to 3 percent and would thus improve air quality only slightly.

If 480 MW of off-peak power were available for hydrogen production, it would be possible to produce enough fuel for 30,000 fleet vehicles, each travelling 100 miles per day, thus accounting for about 5 percent of automobile miles in Phoenix in 2000.

---

*In the U.S. at large, about 8 percent of passenger cars are fleet vehicles.[180] Assuming that the same ratio holds in Phoenix, 30,000 fleet vehicles represents 2 percent of the number of automobiles projected for Phoenix in 2000. But driven 100 miles a day they would account for 5.3 percent of the projected number of automobile miles driven per day.

As PV technology develops and PV hydrogen systems become more competitive, PV sources could augment the off-peak conventional power source in hydrogen production. By the year 2000, PV hydrogen could cost about $13 to $18 per GJ,[186]* about the same as the cost of hydrogen from conventional offpeak power, so that in this timeframe perhaps another 30,000 fleet vehicles could be fuelled with hydrogen, this time from PV sources. The required installed PV generating capacity would be 655 MW, requiring a land area 2.5 square miles for 12-percent efficient PV modules or 1.7 square miles for 18-percent efficient modules.

In the timeframe beyond the year 2000, the use of hydrogen could be extended to the car population at large, provided that these vehicles would have a sufficiently high fuel economy that their range is satisfactory to the general user** and would be equipped with

---

*This may be an overestimate, in part because it is based on the assumption that an entirely new hydrogen system would be constructed when PV technology comes of age. Instead, in this initial phase, it would be possible to use the same electrolyzer at night with off-peak power and during the day with PV power, cutting the capital costs and improving the utilization factor of the electrolyzer. Also, if there were a market for the byproduct oxygen, the cost of hydrogen could be reduced by $1.5 to $2 per GJ.

**Here 50 mpg is assumed, so that the range is 236 km (147 miles). For a car driven 10,000 miles per year (the U.S. average), this implies refuelling every five days, on average. For Phoenix, where the average amount of driving is expected to be over 17,000 miles per car per year in 2015 (see Table 21), the average time between refuellings would have to be every 3 days. At 90 mpg, which is both technically and economically feasible (see Chapter 6), cars in Phoenix would have to refuel every five days. Clearly, higher fuel economy levels would be preferable.

engines having dual- or multi-fuel capability (e.g. stratified-charge engines). If by 2015 half of all cars in Phoenix were fueled with PV hydrogen, the required installed PV generating capacity would be 7.8 GW, and the corresponding land area needed would be 30 (20) square miles for 12-percent (18-percent) efficient PV modules.

A Nationwide Hydrogen Transport System

If initial hydrogen-fuelled cars were successful, fleet vehicles in other areas might be converted to hydrogen. In 1985, there were 7.6 million cars in the United States in fleets of ten or more vehicles.[187] These automobiles averaged about 30,000 miles per year. Assuming an average fuel economy of 50 miles per gallon of gasoline equivalent, the energy consumed by a fleet this size would be about 0.57 EJ per year. If 10 percent of U.S. fleet vehicles converted to PV hydrogen, 0.057 EJ per year of hydrogen production capacity (or 7,000 MW of installed PV capacity) would be needed. At this scale, it would make sense to connect large centralized PV hydrogen facilities in the Southwest to other parts of the country via pipeline.[188] Wider use by the general public would follow once the technology was established for fleet vehicles.

Hydrogen for Residential Space and Water Heating

After large-scale development of transport-based PV hydrogen systems is under way, hydrogen can be considered for use as a fuel for home space and water heating. *(See Chapter 7.)* In much of the Southwest, there is little demand for space heating, and water heating could be provided at a lower cost by PV electricity than by hydrogen. Thus, initial use of PV hydrogen in home heating markets would probably begin in colder regions, using hydrogen transported by pipeline from the Southwest. Significant use of hydrogen for home heating in comparatively cold climates might

follow the establishment of pipelines for transport fuel.

For use in older housing stock, PV hydrogen could be mixed in as a natural gas extender. (In mixtures of up to 10 to 20 percent, natural gas appliances or distribution systems would not have to be modified.) Conversion to pure hydrogen use in existing homes would involve more extensive retrofits of home heating systems, appliances, and local distribution systems. (An analogous retrofit was carried out in the 1930s and 40s when many U.S. cities switched from town gas to natural gas.) However, the best opportunity for introducing hydrogen may be for heating in new super-insulated homes. *(See Chapter 7).*

The successful development of inexpensive home-refuelling systems for hydrogen-powered cars could make the use of hydrogen for space and water heating purposes more attractive. The large hydrogen load required for the automobile would help justify the expense of building a hydrogen distribution network for residences. And instead of visiting a gas station to refuel, the homeowner could simply attach his car to a small compressor in his garage and fill up with hydrogen piped to his home. [While home refuelling of compressed natural gas-powered cars has been proposed, and small compressors are commercially available, additional compressor development is needed to bring down the cost of these systems (now about $4000).[189]]

Parallels with the Development of PV Electricity

As the price of PV modules comes down, a number of markets will open for PV hydrogen, as well as for PV electricity. Table 22 lists some potential markets for PV electricity and PV hydrogen as a function of the PV module price needed to compete with conventional energy sources.

Not surprisingly, most uses of PV systems in the next decade or two are likely to be for elec-

Table 22. Potential Markets for PV Electricity[a] and PV Hydrogen

Electricity Market	Breakeven Solar Module Price ($ per peak Watt)	Potential Market Size Total (MW$_p$)	Hydrogen Market
Corrosion protection	20–100		
Buoys	60		
Consumer products: calculators, etc.	10	100	
Remote water pumping	4–7	2,000	
Diesel generator replacement, remote power	5	10,000	
U.S. utility electric peaking	2–3	50,000	
	1–2	400	Merchant hydrogen
Daytime power for grid connected residences in U.S.	0.7–1.5	100,000	
	0.5–1.0	3,000	Intermediate chemical
	0.20–0.40	4,000	Heating in 2 million super-insulated houses
		7,000	10% of U.S. automotive fleets
		60,000	Large chemical
		600,000	All U.S. home heating
U.S. baseload electric power (with storage)		600,000	
		1,100,000	All U.S. automotive

a. *See Table 4* for assumptions about PV electric markets.

tricity. The first chemical market for PV hydrogen (as a replacement for merchant hydrogen) is likely to open about the same time as the much larger utility peaking market. Rooftop PV systems for residential daytime electricity should become competitive with grid power about the same time that PV hydrogen systems for intermediate-sized chemical users do. When PV modules achieve efficiencies of 12 to 18 percent and costs of $0.2 to $0.4 per peak Watt, however, vast energy markets for baseload

electricity and hydrogen fuel could open. At this stage, the potential markets for PV hydrogen would compare in size to PV electricity markets.

The large-scale development of PV electricity would greatly benefit the development of PV hydrogen, since much of the engineering design of PV systems for central station electric generation will be directly applicable to large PV hydrogen systems.

IX. A Policy Agenda

apid progress being made in PV technology implies that PV hydrogen could begin to be used in energy applications by the turn of the century. But this prospect cannot materialize without new public policy initiatives that would address the environmental and oil-security problems that have prompted interest in clean alternative fuels and speed up the development and demonstration of the technologies needed for a PV hydrogen economy.

Policies for Reducing the External Social Costs of Fossil Fuel Dependency

Since concerns about the risks of fossil fuel dependency are at the root of society's interest in a PV hydrogen economy, policies directed at reducing these risks head the list.

Reducing Local and Regional Air Pollution

In many urban areas of the United States, ozone pollution resulting from emissions from automobiles and gasoline pumps exceeds federally mandated levels. There is also widespread concern about the acid deposition that arises from emissions from power plants and automobiles.

In June of 1989 President Bush announced his Clean Air Plan, which addresses these concerns.[190] The plan calls for reducing volatile

organic compounds (VOCs) emissions significantly by tightening the volatility requirements for gasoline in the summer months and by restricting VOCs emissions when motor vehicles are refueled and operated. Introducing hydrogen as a transport fuel in some polluted urban areas could help meet the goals established in the plan for VOCs. It could also facilitate major reductions in automotive NO_x emissions, which are not targeted in the plan but are important contributors to ozone pollution in some urban areas.

The most far-reaching element of President Bush's plan is a ten-year program to replace part of the motor vehicle fleet in the nine most polluted urban areas with new vehicles that operate on clean-burning fuels. The plan calls for introducing 500,000 clean-fueled vehicles in 1995, 750,000 in 1996, and one million each year from 1997 through 2004. The initial plan does not specify which alternative fuels should be used, but it names methanol, ethanol, and natural gas as possibilities. Because hydrogen contributes much less than these other fuels to air pollution and could become available as a transport fuel in this timeframe, its use should be encouraged in the final Clean Air Plan.

Reducing Oil Import Dependency

U.S. dependency on foreign oil is growing. As a response to low world oil prices, U.S. oil consumption rose 10 percent between 1985 and 1988. In addition, oil production in the lower

48 States has been declining since 1969. Moreover, total oil production has fallen 9 percent since 1985, reflecting the fact that Alaskan oil production has begun to peak. The outlook is that domestic oil production will continue to decline for the indefinite future. As a result, concern about rapidly deepening dependency on imports has led to proposals for a tariff on oil imports[191] or an oil products tax[192] to internalize the societal costs of this dependency. Such a tax on oil would help make alternative fluid fuels more competitive. For perspective, the United States lags behind all other OECD countries in taxing oil. In most other countries gasoline taxes total more than $1 per gallon; in some, they amount to more than $2 per gallon.

Reducing Greenhouse Emissions

An oil tax by itself would help promote fossil synfuels as well as hydrogen. Thus an oil tax should be complemented by a carbon tax that would help limit carbon dioxide emissions, while encouraging both energy efficiency and fuel switching to such renewable sources as PV hydrogen.

If concerns about the greenhouse effect should lead to a decision to make hydrogen costing $15 per GJ at the fuel pump competitive with methanol costing $11 per GJ, a tax of $4 per GJ would be needed. (Producing methanol from coal would release some 37 kg of carbon per GJ of methanol. The carbon tax required to equalize the cost of methanol and PV hydrogen would be about 11 cents per kilogram of carbon.) Adding $4 per GJ to the price of methanol would be equivalent to adding about $2.4 per GJ to the cost of coal feedstock—more than doubling the price of coal for synfuel production. But the impact on the consumer would be much more modest—equivalent to a 50 cents per gallon tax on gasoline.

Creating a Level Playing Field with Tax Policy

Not only does present energy tax policy ignore the external social costs of fossil fuel dependency; it also is biased in favor of fossil fuels. For example, the oil industry is subsidized by such policies as the depletion allowance and the expensing of intangible drilling costs (which would be treated as depreciation in other industries). Such subsidies should be eliminated.

Also, current law is biased against such capital-intensive energy systems as PV electric and PV hydrogen systems,* since under current practice both corporate income taxes and property taxes on energy systems are assessed as capital charges. Since there is no fundamental rationale for this approach to taxation, alternative approaches should be considered. If more than the above-discussed carbon tax on fossil fuels and tax or tariff on oil is needed on energy to meet the federal, state, or local government's revenue requirements, consideration should be given to mechanisms that are not biased against particular fuel forms. One possibility would be a tax proportional to the energy content of the energy carrier.

Promoting Improved Energy End-Use Efficiency

Hydrogen energy systems tend to be both more practical and more economic when the end-use technologies involved are energy-efficient. Thus the broad range of policies needed to implement energy-efficient end-use technologies[193] would help promote the development of the hydrogen economy. These poli-

____
*Since this study is aimed at comparing the costs to society of alternative energy systems, corporate income tax and property taxes were neglected in the analysis. Had these taxes been included as they are presently assessed in the United States, the outlook for capital-intensive energy technologies (PV hydrogen and electricity, hydropower, wind power and nuclear power) would have been less favorable than indicated as compared to fossil fuel-based technologies.

cies would also help society cope with the external social costs of overdependence on fossil fuels before a switch to PV hydrogen can be made.

Since the major initial energy market for PV hydrogen would probably be automotive fuel and since hydrogen-powered vehicles would have to be highly fuel efficient to go several days on a single tank of fuel, improving fuel economy is at the top of the list of energy-conservation measures needed to promote a hydrogen economy.

While an oil tax would help promote fuel-economy improvements, alone it probably would not be a strong enough incentive to raise fuel economy to the high levels needed to develop hydrogen as an automotive fuel. As discussed earlier, the cost per mile of owning and operating a car tends to vary little with fuel price at fuel economies higher than about 30 mpg. *(See Figure 18.)* Thus, consumers driving gasoline-fueled cars would not have a significant economic incentive to seek automotive fuel economy beyond about 30 mpg, even with a substantial gasoline tax. To encourage them to buy more fuel-efficient automobiles, it would be desirable to complement an oil tax with increases in the federally mandated fuel economy standards and/or the levy of "gas guzzler" taxes or similar measures.[194]

Policies Promoting PV Hydrogen Development

While policies to reduce the external social costs of fossil fuel dependency and promote efficient end-use technology are the most important measures needed to facilitate the transition to a hydrogen economy, some initiatives specifically targeting PV hydrogen development are also warranted.

Despite the promise of PV technology, U.S. government support for PV research and development (R&D) has been relatively modest; appropriations for FY 1989 were some $35.5 million—ten percent of the support level for nuclear power and six percent of the support level for fossil fuels (including the Clean Coal Technology program). Not only are these PV support levels less than those for West Germany ($60 million) and Japan ($40–$45 million),[195] but also the Bush Administration's PV budget request for FY 1990 is 29% less than the appropriated budget for FY 1989. At the same time the Administration is seeking a 71% increase in the funding for Clean Coal Technology. *(See Table 23.)*

Research and development on hydrogen energy in the United States is also supported by the government at a relatively modest level of about $3 million per year, compared to $20 million per year in Japan and $50 million per year in West Germany.[196]

Research and Development to Improve the Efficiency and Lifetime of PV Modules and to Reduce Their Manufacturing Cost

Although rapid progress is being made in photovoltaics, increased research and development support is needed for a wide range of PV technologies that hold the promise of meeting the cost and performance goals set forth in this study, with particular emphasis on amorphous silicon, copper indium diselenide, cadmium telluride, and other promising thin-film technologies.

Some specific strategies identified for reducing the manufacturing costs of a-Si solar cells, for example, are the integration of PV module and glass-making facilities, the recovery of silane gas in manufacturing, the development of low-cost frames for large arrays, the automation of production facilities, and the exploitation of economies of scale by expanding a factory's output from present levels of 1 MWp per year to the range 10-100 MWp per year. If all these steps were taken, 12- to 18-percent efficient modules costing $0.2 to $0.4 per peak Watt could probably become available by the year 2000.

Figure 18. The Cost of Driving Versus Automotive Fuel Economy

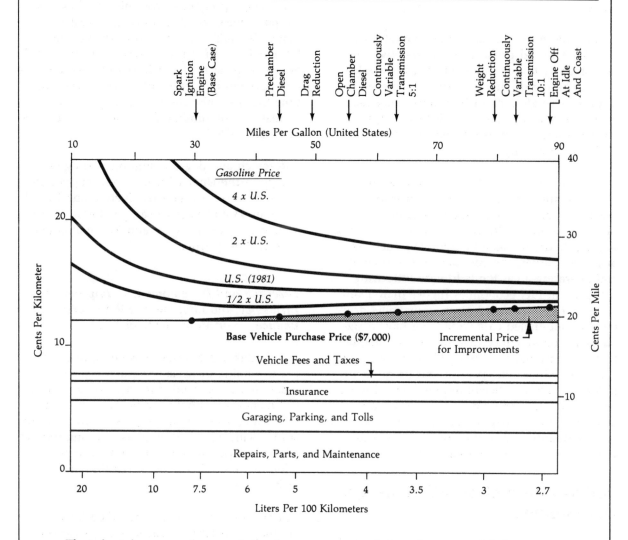

The indicated energy performance is based on computer simulations of an automobile having various fuel economy improvements added in the sequence shown at the top of the graph. The base car is a 1981 Volkswagen Rabbit (gasoline version).

The figure shows that the reduced operating costs associated with various fuel economy improvements are roughly offset by the increased capital costs of these improvements over a wide range of fuel economy.

Source: F. von Hippel and B.G. Levi, "Automotive Fuel Efficiency: The Opportunity and Weakness of Existing Market Incentives," *Resources and Conservation* (1983): 103-124.

Table 23. Department of Energy R&D Budget for Energy Supplies[a]

| | (millions of dollars) | | |
	FY 1988 (actual)	FY 1989 (appropriated)	FY 1990 (requested)
Clean Coal Technology	196.9	190.0	325.0
Fossil Energy	324.1	380.6	163.6
Nuclear	344.7	352.7	350.7
Photovoltaics	34.7	35.5	25.1

a. U.S. Dept. of Energy FY 1990 Congressional Budget Request, DEO/MA-0356, vol. 4, January 1989.

Since the production costs of a-Si solar cells probably vary inversely with the cells' efficiency, achieving a-Si cost targets depends critically on improving the efficiency of commercial a-Si modules from their present values of 5 to 7 percent to the range 12 to 18 percent. Increased support for research and development to improve efficiency and reduce unit capital costs is crucial.

Developing Low-Cost Balance-of-System (BOS) Designs for PV Systems Involving Fixed, Flat-Plate Arrays

The commercialization of PV technology for the utility peaking market in the 1990s should lead to increased efforts to optimize low-cost balance of systems designs for large fixed flat-plate PV arrays. At the PV module costs of $1 to $2 per peak Watt needed for utility PV systems, area-related balance of systems (BOS) costs will still account for only a modest fraction of the overall DC electricity cost. However, as the costs of PV modules drop to the range $0.2 to $0.4 per peak Watt, reduced balance-of-systems costs will become more important. Using estimates drawn from various conceptual designs for large flat-plate PV systems, it has been shown here that area-related BOS costs could be reduced from the often-quoted figure of $50 per m² to perhaps $33 per m². Other analysts have estimated that even lower balance of systems costs may be possible with innovative support structures. For 18-percent efficient PV modules, reducing BOS costs from $50 to $33 per m² corresponds to reducing the PV hydrogen cost by roughly 10 percent or about $1 per GJ. While not as critical as reducing the costs and increasing the efficiencies of the PV modules themselves, reducing BOS costs would clearly help the overall economics of a PV-based hydrogen economy, and efforts to cut these costs should be supported.

Continued Research on Hydrogen End-Use Systems

Among hydrogen-using technologies, hydrogen-powered cars, space-heating and water-heating systems, home car-refuelling systems, and fuel cells warrant high research and development priority, because of both the promising economics and the potentially large markets involved. While some work was done on these technologies in the United States in the 1970s, little is being done today.

Meanwhile, in Germany, Daimler-Benz and BMW are developing hydrogen-powered cars and associated hydrogen storage systems, and research on residential use of hydrogen is under way.

Commercial Demonstrations

Several small (kilowatt-sized) experimental PV hydrogen systems have been built, and a number of conceptual designs for large plants have been developed. But until the recently proposed 600-kW Solar Wasserstoff project in West Germany,[197] there have been no large demonstrations of PV hydrogen systems.

By the mid- to late–1990s, PV hydrogen systems sited in the Southwestern United States could begin to pay their way in markets for chemical hydrogen and oxygen. One or more commercial demonstrations of PV hydrogen systems in the Southwest are needed to encourage chemical companies, energy companies, automobile manufacturers, and/or utilities to exploit these markets. Significantly, the Solar Wasserstoff project is a joint effort involving German automotive, chemical, and aerospace companies, a local utility, and the government.

Commercial demonstrations of hydrogen-powered transport systems are also needed. Such demonstrations should be located in heavily polluted urban centers in the Southwest, where hydrogen could be derived initially from off-peak power from conventional power plants and PV hydrogen could be added to the hydrogen supply near the turn of the century. The transport vehicles involved could be city bus fleets, fleets of delivery vehicles, or fleets of government-owned automobiles.

Conclusion

It would take many decades to complete the transition to a PV hydrogen economy, largely because of capital constraints on the evolution of the energy system.[198] Yet, since the transition could begin in the 1990s, and since the dangers of overdependence on fossil fuels and the benefits of switching to PV hydrogen are so great, policy strategies for hastening the transition warrant high priority.

Joan Ogden is a member of the research staff at the Center for Energy and Environmental Studies at Princeton University. Previously she worked as an applied physicist at the Princeton Plasma Physics Laboratory and at RCA's David Sarnoff Research Center. **Robert H. Williams** is Senior Research Scientist at the Center for Energy and Environmental Studies, Princeton University. In 1988 he received the Szilard Award for Physics in the Public Interest from the American Physical Society "for applying physics to end-use energy efficiency and educating physicists, members of Congress, and the public on energy conservation issues."

Notes and References

(Starred notes are given in more detail in the Technical Version of the Notes and References, which is available as a separate document.)

1. United States Department of Energy, *Energy Security,* A Report to the President of the United States, Washington, DC, March 1987.

2. "Urban Ozone and The Clean Air Act: Problems and Proposals for Change," Staff Paper, Oceans and Environments Program, Office of Technology Assessment, April 1988.

3. *Ibid.*

4. Philip Shabecoff, "Ozone Pollution Is Found at Peak in Summer Heat," *New York Times,* July 31, 1988, p. 1.

5. "Urban Ozone and The Clean Air Act: Problems and Proposals for Change," Office of Technology Assessment.

6. Shabecoff, op. cit.

7. Philip Shabecoff, "Global Warming Has Begun, Expert Tells Senate," *New York Times,* June 24, 1988, p. 1.

8. Don R. Clay, Acting Assistant Administrator for Air, Environmental Protection Agency, Testimony before the Health and the Environment Subcommittee of the House Energy and Commerce Committee, US Congress, February 28, 1989.

9. Michael P. Walsh, "Pollution on Wheels," Report to the American Lung Association, February 11, 1988.

10. *Ibid.*

11. *Ibid.*

12. *Ibid.*

13. "Draft 1988: Air Quality Management Plan," California South Coast Air Quality Management District, Southern California Association of Governments, September 1988.

14. U.S. Environmental Protection Agency, "Guidance on Estimating Motor Emission Reductions from the Use of Alternative Fuels and Fuel Blends," EPA Report No. AA-TSS-PA-87-4, Research Triangle Park, North Carolina, January 1988.

15. C.D. Masters, E.D. Attanasi, W.D. Dietzman, R.F. Meyer, R.W. Mitchell, and D.H. Root, "World Resources of Crude Oil, Natural Gas, Natural Bitumen, and Shale Oil," paper prepared for the 12th World Petroleum Congress, Houston, Texas, 1987.

16. Committee on the Monitoring and Assessment of Trends in Acid Deposition,

National Research Council, *Acid Deposition: Long Term Trends*, National Academy Press, Washington, DC, 1986; J.E. White, ed., *Acid Rain, The Relationship Between Sources and Receptors*, Elsevier, New York, 1988.

17. A.H. Johnson, "Acid Deposition: Trends, Relationships and Effects," *Environment*, v. 28, No. 4, May 1986, p. 6; "Acid Deposition and Forest Decline," Resource Highlights Pamphlet, Land and Water Resources Center, University of Maine, Orono, January 1984.

18. J.J. MacKenzie and M.T. El-Ashry, *Ill Winds: Airborne Pollution's Toll on Trees and Crops*, 71 pp., World Resources Institute, Washington, DC, September 1988.

19. *Ibid.*

20. *Ibid.*

21. "Urban Ozone and The Clean Air Act: Problems and Proposals for Change," Office of Technology Assessment.

22. *Ibid.*

23. B. Bolin, B.R. Doos, J. Jaeger, and R.A. Warrick, *The Greenhouse Effect, Climatic Change, and Ecosystems*, SCOPE 29, John Wiley & Sons, 1986.

24. Philip Shabecoff, "Global Warming Has Begun, Expert Tells Senate," *New York Times*, June 24, 1988, p. 1.

25. W. Haefele *et al.*, *Energy in a Finite World— A Global Systems Analysis*, Ballinger, Cambridge, MA, 1981; World Energy Conference. *Energy 2000–2020: World Prospects and Regional Stresses*, ed. Frisch, J.R., Graham & Trotman, London, 1983; J.A. Edmonds and J. Reilly, "Global Energy Production and Use to the Year 2050," *Energy, the International Journal*, vol. 8, pp. 419–432, 1983; J.A. Edmonds, J. Reilly, J.R. Trabalka, and D.E. Reichle, "An Analysis of Possible Future Atmospheric Retention of Fossil Fuel CO_2," DOE Report No. DOE/OR/21400-1, Washington, DC, 1984.

26. IIASA study by W. Haefele *et al.* (ref. 25)

27. The IIASA low (L) and high (H) projections (ref. 25) are in the box at the bottom of this page.

The carbon emissions (as carbon dioxide) are assumed to be 0.78, 0.63, and 0.43 GT per TW-year for coal, oil, and natural gas respectively. If half of the carbon dioxide released from the use of fossil fuels were to stay in the atmosphere, the atmospheric burden of carbon dioxide would increase from 721 GT (340 ppm) in 1980 to the range 901 to 960 GT (425 to 453 ppm) by 2030, up 55 to 65% from the estimated pre-industrial level of 275 ppm.

28. Bolin (ref. 23)

29. If ΔT_d is the expected equilibrium increase in the surface temperature associated with a doubling of the carbon dioxide concentra-

	Fossil Fuel Use (TW-years/year)			Cumulative Fossil Fuel Use, 1980–2030 (TW-years)		Cumulative Carbon Emissions As CO_2 (Gigatonnes)	
	1980	2030L	2030H	Low	High	Low	High
Coal	2.44	6.45	12.00	206.7	299.7	161.2	233.8
Oil	4.18	5.02	6.83	229.5	269.9	144.6	170.0
N. Gas	1.74	3.47	5.97	125.4	171.3	53.9	73.7
Totals	8.36	14.94	24.80	561.6	740.9	359.7	477.5

tion, then the equilibrium temperature increase ΔT resulting from an X-fold increase in the CO_2 level is given by

$$\Delta T = \Delta T_d \times (\ln X) / (\ln 2).$$

For the IIASA scenarios, $X = 1.55$ to 1.65 by 2030 (see note 27), so that

$$\Delta T = (0.63 \text{ to } 0.72) \times \Delta T_d \text{ by } 2030.$$

30. Bolin (ref. 23)

31. Bolin (ref. 23)

32. R.H. Williams, E.D. Larson, and M.H. Ross, ''Materials, Affluence, and Industrial Energy Use,'' *Annual Review of Energy*, vol. 12, pp. 99–144, 1987; Energy Demand and Materials Flows in the Economy,'' *Energy*, vol. 12, nos, 10/11, pp. 953–967, 1987; E.D. Larson, M.H. Ross, and R.H. Williams, ''Beyond the Era of Materials,'' *Scientific American*, vol. 254, no. 6, pp. 34–41, June 1986.

33. J. Goldemberg, T.B. Johansson, A.K.N. Reddy, and R.H. Williams, *Energy for a Sustainable World*, 119 pp., World Resources Institute, Washington, DC, September 1987; J. Goldemberg, T.B. Johansson, A.K.N. Reddy, and R.H. Williams, *Energy for a Sustainable World*, 517 pp., Wiley-Eastern, Delhi, 1988; J. Goldemberg, T.B. Johansson, A.K.N. Reddy, and R.H. Williams, *Energy for Development*, 73 pp., World Resources Institute, Washington, DC, September 1987.

34. R.H. Williams, ''Are Runaway Energy Capital Costs a Constraint on Development?'', paper presented at the International Seminar on the New Era in the World Economy, Fernand Braudel Institute of World Economics, São Paulo, Brazil, August 31– September 2, 1988.

35. Fossil fuel use for the global energy future described in *Energy for a Sustainable World* (ref. 33) is as follows:

	Fossil Fuel Use (TW-years/ year) 1980	2020	Cumulative Fossil Fuel Use, 1980–2020 (TW-years)	Cumulative Carbon Emissions As CO_2 (Gigatonnes)
Coal	2.44	1.94	87.6	68.3
Oil	4.18	3.21	147.8	93.1
N. Gas	1.74	3.21	99.0	42.6
Totals	8.36	8.36	334.4	204.0

If half of the carbon dioxide released from the use of fossil fuels were to stay in the atmosphere, the atmospheric burden of carbon dioxide would increase from 721 GT (340 ppm) in 1980 to 823 GT (388 ppm) by 2020, up 41% from the estimated pre-industrial level of 275 ppm, and the equilibrium global warming from this much added carbon dioxide would be:

$$\Delta T = \Delta T_d \times (\ln 1.41) / (\ln 2) = 0.5 \times \Delta T_d.$$

36. Conference Statement of the World Conference on The Changing Atmosphere: Implications for Global Security, Toronto, June 27–30, 1988.

37. J.B.S. Haldane was the first to suggest electrolytic production of hydrogen fuel from wind power in 1923, and in 1927, A.J. Stuart suggested using hydropower as a source of electricity for hydrogen production. The use of solar power to produce hydrogen was first suggested by J. O'M. Bockris in 1962. (See J. O'M. Bockris, *Energy Options*, Halsted Press, New York, 1980.)

38. See Appendix A in *Energy for a Sustainable World*, Wiley-Eastern, Delhi, 1988 (ref. 33).

39. H.K. Schneider and W. Schulz, ''Investment Requirements of the World Energy Industries 1980–2000,'' Report prepared for the World Energy Conference Study Group on Long-Term Investment Requirements: Needs, Constraints, and Proposals, World

Energy Conference, London, September 1987.

40. See Appendix A in *Energy for a Sustainable World*, Wiley-Eastern, Delhi, 1988 (ref. 33).

41. When manufacturing single crystal silicon solar cell material, the atoms must be arranged in a precise geometrical "lattice." This is done by inserting a small "seed crystal" into purified molten silicon, then withdrawing and cooling it slowly to preserve the original pattern of the seed. One common method involves producing a cylindrical crystalline ingot, several inches in diameter, which is then sawn into circular wafers. The circular wafers are then positioned and electrically connected to make a solar module. This method is costly, as about half of the material is lost in sawing, and it is time consuming to assemble the modules. A potentially less expensive method of single crystal solar cell mass production involves pulling a continuous rectangular sheet or "web" of single crystal silicon. Material and energy consumption are reduced with this technique.

42. Crystalline silicon solar cells were originally developed as power sources for spacecraft and satellites. In the early 1970's, the cost of cells for space applications was about $120 per peak Watt (in 1986$). During the energy crisis of the mid-1970's, there was growing interest in terrestrial use of photovoltaics, and less expensive manufacturing techniques were developed. As of 1974, solar cells cost about $50 per peak Watt; and by 1976, the cost of solar cells had dropped to about $25 per peak Watt. See D.L. Pulfrey, *Photovoltaic Power Production*, Van Nostrand Reinhold, New York, 1978.

43. Costs for solar modules are given in $/peak Watt ($/Wp) or sometimes in dollars per square meter ($/m²) or dollars per square foot. Peak Wattage is the amount of electrical power generated under maximum insolation (sunlight). Under ideal conditions the

earth's surface receives a maximum of about 1 kilowatt of solar power per square meter. The cost per unit area and the cost per unit of output are related by:

$$\$/Wp = (\$/m^2)/[(\text{efficiency}) \times (1000 \ Wp/m^2)]$$

Production volumes for solar cells are generally given in peak Watts of PV generating capacity produced per year or "Wp/year."

44. In this study the real (inflation-corrected) discount rate is assumed to be 6.1% for all energy production systems and the insurance rate is assumed to be 0.5% per year [values recommended by the Electric Power Research Institute for evaluating alternative electric utility investments (Electric Power Research Institute, *Technical Assessment Guide, Vol. 1: Electricity Supply—1986*, EPRI P-4463-SR, December 1986—henceforth referred to as the 1986 EPRI TAG)]. Corporate income taxes and tax incentives are neglected. For consumer systems such as automobiles and home heating systems, a discount rate of 10% is assumed.

*45. The cost of electricity produced in the southwestern U.S. from solar cells costing $120/Wp would be about $5.12/kWhDC, assuming that the PV system lasts 30 years and neglecting indirect costs, balance of system costs, and operation and maintenance costs.

*46. The following are estimates of the costs of producing electricity from new coal or nuclear plants in the United States:

Type	Coal(a)	LWR(b)
Unit Size [MW(e)]	500	1100
Efficiency (Fuel-to-Busbar)	34.6	33.4
Unit Capital Cost (1986$/kW)	1375	2990

Levelized Busbar Cost (1986$/kWh)

Capital (c)	0.0189	0.0411
Fuel (d)	0.0181	0.0087
O&M (e)	0.0101	0.0105
TOTAL	0.0471	0.0604

a. For a conventional coal-fired subcritical steam plant with wet limestone flue gas desulfurization, for bituminous coal, East/West Central regions. Unit capital costs, efficiencies, O&M costs, and other plant characteristics are from the 1986 EPRI TAG.

b. Unit capital costs, efficiencies, and other light water reactor plant characteristics are from the 1986 EPRI TAG.

c. For a 30 year plant life (so that CRF + INS = 0.0784) and a 65% capacity factor.

d. For a coal price of $1.74/GJ—the average price projected by the U.S. Department of Energy for the year 2000 (Energy Information Administration, "Annual Energy Outlook 1987, with Projections to 2000," DOE/EIA-0383(87), March 28, 1988). For a nuclear fuel cycle cost of $0.81/GJ, the estimated cost in 1990, according to the EPRI TAG.

e. The O&M cost estimate for coal plants is from the 1986 EPRI TAG. That for nuclear plants is the actual US average for nuclear plants in the period 1982–1985, according to Energy Information Administration, "Historical Plant Cost and Anuual Production Expenses for Selected Electric Plants 1985," DOE/EIA-0455(85).

47. H.M. Hubbard, "Photovoltaics Today and Tomorrow," *Science*, v. 244, pp. 297–394, April 21, 1989; Robert Poole, "Solar Cells Turn 30," *Science*, v. 241, pp. 900–901, August 19, 1988; Taylor Moore, "Opening the Door for Utility Photovoltaics," *EPRI Journal*, January/February 1987.

48. Taylor Moore, "Opening the Door for Utility Photovoltaics," *EPRI Journal*, January/February 1987.

49. D.E. Carlson, "Low-Cost Power from Thin-Film Photovoltaics," in *Electricity: Efficient End-Use and New Generation Technologies and Their Planning Implications*, T.B. Johansson, B. Bodlund and R.H. Williams, eds., Lund University Press, Lund Sweden, 1989; Z Erol Smith III, "Amorphous Silicon for Solar Cell Applications: Defect Generation, Annealing, and Equilibrium," Department of Electrical Engineering, Princeton University, 1987.

50. H.M. Hubbard, "Photovoltaics Today and Tomorrow," *Science*, v. 244, pp. 297–304, April 21, 1989; K. Zweibel, H.S. Ullal, R.L. Mitchell. "The DOE/SERI Polycrystalline Thin-Film Subcontract Program," Proceedings of the 20th IEEE PV Specialists Conference, Las Vegas, Nevada, 1988; "Special Issue on Cadmium Telluride," Solar Cells, v. 23, No. 1–2, January/February 1988.

51. Amorphous silicon is vapor deposited from a silicon-bearing gas (silane) on a glass, plastic or stainless steel substrate in a vacuum chamber. The depth and composition of the various layers which make up the solar cell are controlled by adjusting the composition of the gas in the deposition chamber. Solar modules can be configured swiftly and flexibly from these sheets of amorphous silicon. The necessary electrical connections are made by scribing the amorphous silicon film with a laser and depositing metal conducting paths.

52. D.E. Carlson, "Low-Cost Power from Thin-Film Photovoltaics," in *Electricity: Efficient End-Use and New Generation Technologies and Their Planning Implications*, T.B. Johansson, B. Bodlund and R.H. Williams, eds., Lund University Press, Lund Sweden, 1989.

53. *Photovoltaic Insiders' Report*, February 1988.

54. S. Kaplan, Chronar Corp., private communications, 1988.

55. M.L. Wald, "Solar Power Plant Planned for California," Business Day Section of *The New York Times*, September 6, 1988.

56. D.E. Carlson, General Manager, Thin Film Division, Solarex, private communication, 1987.

57. *Photovoltaic Insiders' Report*, January 1987.

58. S. Kaplan, Chronar Corp., private communication, 1988.

59. Arun Madan, "Amorphous Silicon: from Promise to Practice," *IEEE Spectrum*, pp. 38–43, September 1986.

60. M.L. Wald, *op. cit.*

61. Photovoltaic power may prove to be attractive to many developing countries. Not only is sunshine a plentiful resource, but the technology can be readily acquired, owned and operated. Although the manufacture of solar modules requires a factory using a "high-tech" process, the installation of flat plate solar modules requires little specialized knowledge beyond basic construction skills. Experience with PV projects in the U.S. and abroad have shown that maintenance needed on flat plate solar arrays is simple and minimal.

In the last year or two China, as well as France and Yugoslavia, decided to get into the amorphous silicon manufacturing business, and one U.S. company (Chronar Corporation) specializes in selling "turnkey" amorphous silicon solar cell factories to interested countries. A plant with a production capacity of 1–1.4 MW peak/year costs about $6–$7.5 million, a relatively small initial investment for getting into a new energy supply business. (Source: S. Kaplan, Chronar Corporation, private communications, 1988.)

62. Photovoltaic power is well-suited for utility peaking service, as the maximum output of a solar array generally occurs at the same time as the maximum electricity demand for air conditioning, on hot, sunny summer days.

63. If the estimated two million remote villages in the world were electrified with 5 kW photovoltaic plants as an alternative to using small diesel generators, the annual oil savings would be about 0.2 EJ or 0.7% of total oil use by developing countries in 1986.

Installing 40 GW of photovoltaic peaking capacity in the U.S. (equal to the total oil and natural gas fired peaking capacity) could displace up to 1.1 EJ of oil and gas now used for power generation, some 2% of total oil and gas use in the U.S.

If 50 million residential users in the U.S. bought 2 kW rooftop PV generators to produce daytime electricity, the electricity produced could displace 2.7 EJ of oil and gas now used for electric power generation, or 5% of total U.S. oil and gas use in 1986. [These savings and the peaking power savings indicated above are not additive, as daytime residential electricity demand is usually a major contributor to the utility system peak demand.]

64. One new technology under development (Stuart Licht, MIT) is the liquid solar cell that combines the functions of converting solar energy into electricity and storage in a single system. See, for example, "Solar Cells, Mehr Licht," The *Economist*, pp. 82–83, January 16–22, 1988.

*65. As PV electricity is generated only when the sun shines, a method of electricity storage must be found to provide PV electricity for other times. In general, the amount of storage needed depends in complicated ways on the details of the variations of the output of the PV array and of the profile of the electrical load that must be met by the PV system. To give an indication of the potential significance of storage costs, a simple model is constructed here that estimates the cost of baseload electricity from a PV + storage system located in a sunny area, such as the southwestern United States. While this model overestimates the amount of storage that would be needed in practical applications (since in most situations electricity demand tends to be much higher during the period of the day when insolation is the greatest than at other times), this approach permits a ready comparison between PV and alternative systems that can provide baseload electricity.

This model yields the following costs for baseload electricity (in $/kWh AC) for several alternative storage technologies:

Storage Technology

For DC input PV electricity @	$0.020/ kWh	$0.035/ kWh
Lead-acid batteries	0.074	0.093
Advanced batteries	0.047	0.065
Underground pumped storage	0.055	0.074
Hydrogen/air turbine	0.072	0.104
Hydrogen/oxygen turbine	0.056	0.081

Comparing these costs to the estimated costs of baseload electricity from new coal and nuclear plants in the U.S. (note 46), it is seen that in sunny areas PV power plus advanced batteries would be competitive with power produced in new coal and nuclear plants for PV DC electricity costs of $0.02 per kWh or less and with current nuclear power for a DC electricity cost of $0.031 per kWh. Costs are based on storage technology parameters from the EPRI TAG, 1986.

66. For detailed information about the uses of hydrogen see, for example:

K.E. Cox and K.D. Williamson, *Hydrogen Its Technology and Implications*, five volumes, CRC Press, Boca Raton, Florida, 1979.

T.N. Veziroglu, *Hydrogen Energy: Parts A and B*, 2 volumes, Plenum Press, New York, 1975.

The International Journal of Hydrogen Energy, started in 1976, is the primary publication of hydrogen energy research community.

See also reports from the Brookhaven National Laboratory hydrogen program (mid-1970s to 1985).

T.N. Veziroglu, ed., *Hydrogen Energy Progress*, Proceedings of the World Hydrogen Energy Conferences, Pergamon Press, 1974–1988.

67. Private communications from Sigurd Wagner, Electrical Engineering Department, Princeton University, July 1988.

68. Four US manufacturers (Chronar, Solarex, Arco Solar, and ECD) are in a cost shared program with the US Department of Energy to produce modules of these efficiencies by 1990, *PVIR*, February 1987.

69. D.E. Carlson, Solarex Thin Films Division, private communications, 1988; D.E. Carlson, "Low-Cost Power from Thin-Film Photovoltaics," in *Electricity: Efficient End-Use and New Generation Technologies and Their Planning Implications*, T.B. Johansson, B. Bodlund and R.H. Williams, eds., Lund University Press, Lund, Sweden, 1989.

70. E.A. DeMeo and R.W. Taylor, "Solar Photovoltaic Power Systems: An Electric Utility Perspective," *Science*, v. 224, April 20, 1984; H.M. Hubbard, "Photovoltaics Today and

Tomorrow," *Science*, v. 244, pp. 297–304, April 21, 1989.

71. Private communications from Sigurd Wagner, Electrical Engineering Department, Princeton University, July 1988.

72. D.L. Staebler and C.R. Wronski, *Applied Physics Letters*, v. 31, p. 292, 1977.

73. D.E. Carlson, "Solar Cells," in *Semiconductors and Semimetals*, v. 21, Part D, ed. J.I. Pankove, Academic Press, NY, p. 7, 1984.

74. D.E. Carlson, 8th European Photovoltaic Solar Energy Conference, Florence, Italy, May 9–13, 1988.

75. U.S. Department of Energy, DOE/CH10093-19, January 1988. Also EPRI, Sandia reports.

76. S. Kaplan, Chronar Corp., private communications, 1988.

77. "$1 per Wp Module Cost Target Seen Obtainable by Early 1990s Without Efficiency Gains," *PVIR*, p. 4, May 1988.

78. M.L. Wald, *op. cit.*

79. "Chronar Negotiating to Install 60 MW, $150 Million PV System in Southern California," *PVIR*, p. 1, September 1988.

80. D.E. Carlson, "Low-Cost Power from Thin-Film Photovoltaics," in *Electricity: Efficient End-Use and New Generation Technologies and Their Planning Implications*, T.B. Johansson, B. Bodlund and R.H. Williams, eds., Lund University Press, Lund, Sweden, 1989.

81. Private communications from Sigurd Wagner, Electrical Engineering Department, Princeton University, July 1988.

82. D.E. Carlson, General Manager, Thin Film Division, Solarex, private communications, 1988.

83. S.L. Levy and L.E. Stoddard, "Integrated Photovoltaic Central Station Conceptual Designs," EPRI Report AP-3264, June 1984.

84. G.T. Noel, D.C. Carmichael, R.W. Smith, and J.H. Broehl, "Optimization and Modularity Studies for Large-Size, Flat-Panel Array Fields," Battelle-Columbus, 18th IEEE PV Specialists' Conference, Las Vegas, Nevada, October 1985.

85. *See note 89.*

86. Hydrogen transmission costs can be less than those of synthetic medium BTU coal gas and comparable to those for natural gas, with optimized pipelines. For a 1000 mile (1600 km) pipeline, the cost of hydrogen transmission is about $0.35/GJ. See note 100 and D. Christodoulou, "Technology and Economics of the Transmission of Gaseous Fuels Via Pipeline," Princeton University, Department of Mechanical and Aerospace Engineering, Master's Thesis, 1984.

The cost of long-distance DC electrical transmission is about $0.09 to $0.11 per GJ per 100 km for distances longer than about 1000 km. For a 1600 km transmission line, this is a cost of about $1.44 to $1.76/GJ, roughly four times the cost of hydrogen transmission via pipeline. See D.W. Devins, *Energy : Its Physical Impact on the Environment*, John Wiley and Sons, New York, 1982.

*87. The interface between the PV array and the electrolyzer must be considered when calculating the efficiency of a PV hydrogen system. It is possible to achieve about 93% coupling efficiency in experimental PV hydrogen systems. Since both PV arrays and electrolyzers are modular, a good match should be achievable for any desired voltage/current combination.

C. Carpetis, *International Journal of Hydrogen Energy*, Vol. 7, p. 287, 1982; C. Carpetis, *International Journal of Hydrogen Energy*, Vol. 9, p. 969, 1984; R.W. Leigh, P.D. Metz, and K. Michalek, "Photovoltaic Electrolyzer System Transient Simulation Results," BNL-34081, December 1983; P.D. Metz and M. Piraino, BNL-51940, July 1985.

88. We have considered only fixed, flat-plate arrays, because the experience gained from PV projects to date shows that gains in efficiency possible with tracking and/or concentrating systems are not sufficient to offset the added complexity of construction and maintenance.

    Sources: G.T. Noel, D.C. Carmichael, R.W. Smith, and J.H. Broehl, "Optimization and Modularity Study for Large-Size PV Flat-Panel Array Fields," Batelle-Columbus, 18th IEEE PV Specialists Conference, Las Vegas, Nevada, October 1985; G.J. Shushnar, J.H. Caldwell, R.F. Reinoehl, and J.H. Wilson, "ARCO Solar Field Data for Large PV Arrays," 18th IEEE PV Specialists Conference, Las Vegas, October 1985.

*89. Balance of systems costs for amorphous silicon solar cell flat-plate fixed PV arrays are discussed in this note. Combining the best features of several conceptual design studies cited in Table 6, we found that it was possible to reduce area-related BOS costs to $33/m^2.

90. G. Grassi, P. Paoli, L. Leonardini, E. Vitali, P. Conti, E. Colpizzi, "Low-Cost Support Structures for Large Photovoltaic Generators," 18th IEEE Photovoltaic Specialists' Conference, Las Vegas, Nevada, October 1985; T.B. Taylor and H.B. Warren, "The Development of Low Cost Photovoltaic Systems Based on Amorphous Silicon Solar Cells," NOVA, Inc., draft, October 1984.

91. The system efficiency is defined as the electrical energy produced by the entire PV array divided by the incident solar energy. System efficiency includes losses due to wiring of modules, etc. Typically a 10% efficient system is made up of 12% efficient modules; a 15% efficient system of 18% efficient modules.

*92. The cost of producing DC photovoltaic electricity from a large (> 5 MW) array is calculated in this note and shown in Figure 8, for El Paso, Texas.

93. There are currently three electrolysis technologies commercially available or under development. Two of these, the "unipolar" and "bipolar" alkaline electrolyzers, which form hydrogen and oxygen by passing direct current through an aqueous solution of potassium hydroxide (lye), are commercially available, mature technologies. Only modest refinements and cost improvements are expected in the next 5-15 years. Unipolar electrolyzers are less expensive, more efficient, and more modular; bipolar electrolyzers have the advantage of high pressure operation, which saves on compression costs for the produced hydrogen. The third type of electrolyzer, which uses solid polymer electrolytes (SPE) made of an acid resin, is still in the demonstration stage. At present, platinum catalysts are required for stable operation. Although the search for less expensive and rare catalyst materials continues, a long-term solution to this problem is uncertain. In the present study only alkaline type electrolyzers are considered.

    It has been suggested that running the electrolyzer at high pressure could reduce overall hydrogen costs by eliminating the need for external compression. The tradeoffs between less costly, more efficient unipolar electrolyzers and bipolar electrolyzers that require less compression of the produced hydrogen are considered for various applications in the present study.

*94. The cost of PV powered electrolytic hydrogen production is estimated in this note.

Sources for costs and performance of electrolyzers: R.L. Leroy and A.K. Stuart, ''Unipolar Water Electrolyzers: a Competitive Technology,'' in *Hydrogen Energy System*, Proceedings of the 2nd World Hydrogen Energy Conference, Zurich, 21–24 August, 1978; private communication from R.F. Craft, Electrolyser Corporation, Ontario, Canada, 1985; E. Fein and K. Edwards, ''Market Potential of Electrolytic Hydrogen Production in Three Northeastern Utilities' Services Territories,'' Electric Power Research Institute Report EPRI EM-3561, May 1984.

*95. In this note the efficiency and capital cost of electrolyzers are described as a function of hydrogen output capacity. *Sources:* E. Fein and K. Edwards, ''Market Potential of Electrolytic Hydrogen Production in Three Northeastern Utilities' Service Territories,'' EPRI Report, EM-3561, May 1984; R.F. Craft, Electrolyser Corp., private communications, 1985.

96. Most design studies for central station PV power plants have chosen 1-10 MW as an appropriate ''building block'' size. There is little economy of scale above sizes of about 5-10 MW (G.T. Noel, D.C. Carmichael, R.W. Smith, and J.H. Broehl, ''Optimization and Modularity Study for Large-Size Photovoltaic Flat-Panel Array Fields,'' 18th IEEE Photovoltaic Specialists' Conference, Las Vegas, Nevada, October 1985).

97. Oxygen is produced in electrolysis at a rate of 0.0563 tonnes per GJ of produced hydrogen. The net value of the oxygen byproduct is estimated to be $1.5–$2.2 per

GJ for hydrogen (in 1986 US$). (See note 162, and M. Hammerli, ''When Will Electrolytic Hydrogen Become Competitive?'' *Journal of Hydrogen Energy*, vol. 9, no. 1/2, pp. 25–51, 1984.)

*98. In this note, alternative types of hydrogen storage are described and the cost of hydrogen storage is estimated.

1. *Large-Scale Hydrogen Storage for an Energy System:* Gaseous hydrogen can be stored underground in depleted oil or gas fields, aquifers, and rock or salt caverns. Such formations typically have very large capacities (up to 10^9 Nm³, or large enough to store up to 0.01 EJ of hydrogen). Underground storage costs would range from $0.19/GJ for gas wells to $0.70/GJ for rock or wet salt caverns, assuming that electricity cost $0.02/kWh.

	Rock Cavern	Salt Cavern ''Wet''	Salt Cavern ''Dry''	Gas Well
Capital + Install	0.348	0.344	0.305	0.131
Power *	0.132	0.147	0.132	0.022
O+M	0.204	0.206	0.186	0.035
Total	0.685	0.698	0.624	0.188

*For electricity costing $0.020/kWh.

Source: J.B. Taylor, J.E.A. Alderson, K.M. Kalyanam, A.B. Lyle, and L.A. Phillips, ''Technical and Economic Assessment of Methods for the Storage of Large Quantities of Hydrogen,'' *International Journal of Hydrogen Energy*, Vol. 11, p. 5, 1986.

2. *Storage of Gaseous Hydrogen for Industrial Use:* Alternatively, gaseous hydrogen can be stored in rows of pressurized steel tanks above ground typically at 450-2500 psi. Tank storage is modular, with very little economy of scale. Most industrial hydrogen is currently stored this way in the U.S. At the present, the total U.S. storage capacity is about 10 days hydrogen

production, or about 3% of the annual total.

For small-scale industrial gaseous hydrogen storage, pressurized steel tanks are the best alternative, unless liquid hydrogen is needed at the point of end use. In a survey by Fein and Edwards small hydrogen users reported storage costs between $1.5 and $15/GJ depending on the application.

Source: E. Fein and K. Edwards, "Market Potential of Electrolytic Hydrogen Production in Three Northeastern Utilities' Service Territories," EPRI Report, EM-3561, May 1984.

*99. The cost of hydrogen compression (in $/GJ) in a large system is estimated in this note. The results are:

Summary of Costs for Hydrogen Compression ($/GJ)

P1→P2(psia)	Capital	Power	Total (for Ce = $0.02/kWh)
14.7→450	0.69	15.1 × Ce	0.99
450→1000	0.15	3.33 × Ce	0.22
14.7→1000	0.84	18.4 × Ce	1.21

where:

P1 = initial pressure (electrolyzer pressure)
P2 = final pressure (pressure at pipeline inlet)
Ce = cost of electricity to run compressor ($/kWh)

*100. In this note the cost of hydrogen transmission via long distance pipeline is estimated, based on D. Christodoulu, *Technology and Economics of the Transmission of Gaseous Fuels via Pipelines*, Thesis, Master of Science in Engineering, Princeton University, April, 1984.

For hydrogen transmission of 1000 miles, the optimum pipe is about 63 inches in diameter, the pipeline costs about $1050 per linear meter, and the transmission cost is $0.35/GJ.

101. J. Pangborn and M.I. Scott, "Domestic Uses of Hydrogen," in *Hydrogen: Its Technology and Implications*, D.A. Mathis, ed., Energy Technology Review No. 9, Noyes Data Corporation, 1976.

102. Diomedes Christodoulu, *Technology and Economics of the Transmission of Gaseous Fuels via Pipelines*, Thesis, Master of Science in Engineering, Princeton University, April, 1984.

103. Hydrogen from refinery stack gases has been used as a natural gas extender by some utilities. In addition, some synthetic natural gases from petroleum products are naturally rich in hydrogen. In Honolulu, manufactured gas typically has a hydrogen content (by volume) of 10%. No adjustment of end-use systems is necessary at this modest level of hydrogen blending.

Source: W.J.D. Escher, R.W. Foster, R.R. Tison, and J.A. Hanson, "Solar/Hydrogen Systems Assessment," DOE/JPL-9559492, 1980.

104. Birgir Arnason, *Methanol from Biomass and Urban Refuse: Prospects and Opportunities*, Thesis, Master of Science in Engineering, Princeton University, 1983.

105. Conference Statement of the 1988 World Conference on The Changing Atmosphere: Implications for Global Security, *op. cit.*

106. R.K. Lester, "Rethinking Nuclear Power," *Scientific American*, vol. 254, no. 1, pp. 31–39, March 1986; I. Spiewak and A.M. Weinberg, "Inherently Safe Reactors," *Annual Review of Energy*, vol. 10, pp. 431–462, 1985; J.J. Taylor, "Improved and Safer Nuclear Power," *Science*, vol. 244, pp. 318–325, 21 April 1989; J.F. Ahearne, "Will Nuclear Power Recover in a Greenhouse?" Discussion Paper ENR89–06, Resources for the Future, Washington, DC, May 1989.

107. D. Albright and H. Feiveson, "Why Plutonium Recycle?" *Science*, vol. 235, pp. 1555–1556, 27 March 1987; D. Albright and H. Feiveson, "Plutonium Recycling and the Problem of Nuclear Weapons Proliferation," *Annual Review of Energy*, vol. 13, pp. 239–265, 1988.

108. In a full-blown plutonium economy, involving plutonium breeder reactors and light water reactors operated on closed fuel cycles, spent fuel would be reprocessed to recover the plutonium and recycle it in fresh fuel. The plutonium discharge rate would be about 290 kg per billion kWh for a liquid metal fast breeder reactor and 193 kg for a pressurized water reactor fueled with plutonium and natural uranium (T.H. Pigford and C.S. Yang, "Thorium Fuel-Cycle Alternatives," Report EPA-520/6-78-008, prepared for the U.S. Environmental Protection Agency, November 1978). In equilibrium each 1 GW of breeder capacity could produce enough fuel to meet its own recurring needs and support 0.42 GW of light water reactor capacity (H.A. Feiveson, F. von Hippel, and R.H. Williams, "Fission Power: an Evolutionary Strategy," *Science*, vol. 203, pp. 330–337, January 26, 1979). Thus in a full-blown

plutonium economy the plutonium discharge rate would average about 261 kg per billion kWh—or 1487 kg per installed GW(e), for nuclear plants operating at a 65% average capacity factor.

The global consumption of fossil fuels in 1986 was as follows (British Petroleum Company, "BP Statistical Review of World Energy," London, June, 1987):

oil	2881.0 MTOE	=	120.6 EJ
natural gas	1507.1 MTOE	=	63.1 EJ
coal	2309.1 MTOE	=	96.7 EJ
Total	6697.2 MTOE	=	280.4 EJ

Replacing 1/4 of coal with nuclear electricity would require 2.22 trillion kWh of nuclear electricity (assuming that nuclear power displaces coal-based power plants that are 33% efficient). Replacing 1/4 of oil and gas with hydrogen derived via electrolysis would require some 15.84 trillion kWh of nuclear electricity (assuming a rectifier efficiency of 96% and an electrolyzer efficiency of 84%). Thus displacing 1/4 of fossil fuels would require producing annually some 18.06 trillion kWh of nuclear electricity (compared to total world electricity production of 9.27 trillion kWh in 1984). At an average capacity factor of 65% this would require 3200 GW of installed nuclear generating capacity and the annual discharge in spent fuel of some 4.7 million kg of plutonium.

109. R.H. Williams and H.A. Feiveson, "Diversion-Resistance Criteria for Future Nuclear Power," PU/CEES Report No. 239, Center for Energy and Environmental Studies, Princeton University, June 1989.

110. See Appendix A in J. Goldemberg, T.B. Johansson, A.K.N. Reddy, and R.H. Williams, *Energy for a Sustainable World*, Wiley-Eastern, 1988. See note 33.

111. *Ibid.*

112. R.H. Williams, ''Potential Roles for Bio-energy in an Energy-Efficient World,'' *Ambio*, vol. 14, nos. 405, pp. 201–209, 1985.

113. *Ibid.*

*114. In this note the costs of synthetic fuel production from coal and biomass are estimated.

115. The annual water consumption in the U.S. is about 144×10^{12} liters per year. The annual petroleum consumption is 41 EJ or 1.24×10^{12} liters per year. The ratio of water to petroleum use (on a volumetric basis) is then $144/1.24 = 116$. See J. Harte *Consider a Spherical Cow*, William Kaufmann, Inc., Los Altos, California, 1985.

116. Consider hydrogen production in a desert area with average insolation like that near El Paso (271 Watts per square meter on a collector tilted at the latitude angle). As the collectors are tilted, the rows of collectors must be spaced to prevent shading in the winter when the sun is low in the sky. To prevent all shading until 3 pm at the time of the winter solstice, the ground area must be 2.04 times the collector area. For a 15% efficient PV system and 84% efficient electrolytic units, the hydrogen energy production rate per square meter of land area would be:

$0.84 \times 0.15 \times (271 \text{ W/m}^2) \times (3.1536 \times 10^7$ seconds/year)$/2.04 = 0.53$ GJ/m²/year.

Since the consumptive water requirements for electrolysis are 63 liters per GJ of produced hydrogen, (the water required to cool the electrolyzer can be recirculated), the rainfall required to support this consumptive water requirement is just 3.4 centimeters per year. For a 10 percent efficient PV system only 2.3 centimeters of rainfall would be required.

*117. Consider first synthetic liquids from coal. The total amount of U.S. land which could be strip mined for coal is 24,000 square miles (62,000 square kilometers). If all this land were strip mined, it would produce an estimated 86 billion metric tonnes of coal. Assuming an average heating value of 20 GJ/tonne for the coal, 10 years for strip-mined land reclamation, and a 55% average efficiency for converting coal into a synthetic liquid fuel *(see note 114)* the amount of synthetic liquid fuel produced per unit of land area used would be:

$(86 \times 10^9 \text{ tonnes}) \times (20 \text{ GJ/tonne}) \times 0.55 / (6.2 \times 10^{10} \text{ m}^2 \times 10 \text{ years}) = 1.52$ GJ/m²/year.

Next consider the production of PV hydrogen. A 15% efficient PV system connected to an 84% efficient electrolyzer would produce about 0.53 GJ/m²/year (assuming that land equal to 2.04 times the collector area is needed to avoid shading the array). Thus, according to this estimate, about one third as much land would be required for coal synfuels as for PV hydrogen production.

In the above comparison, the estimate of coal resources is an average over all U.S. land which could potentially be strip mined. This would include some regions with a low coal yield and may overestimate the amount of land needed for more productive near-term coal fields.

Source: G. Atwood, ''The Strip-Mining of Western Coal,'' *Scientific American*, December 1975.

Now consider synthetic liquids from biomass. Assuming an average insolation rate of 200 W/m² in biomass-producing areas, an average conversion efficiency of sunlight to biomass of 1/2%, and an average conversion efficiency of 60% for producing synthetic fuel (see note 114), the annual synthetic fuel production rate from biomass would be:

$$0.005 \times (200\ W/m^2) \times (3.1536 \times 10^7$$
$$seconds/year) \times 0.60 = 0.0189\ GJ/m^2/year$$

The amount of land needed for biomass synfuel production in this case would be 28 times that required for production of PV hydrogen in the Southwest.

118. J.C. Fisher, *Energy Crises in Perspective.* Wiley, New York, 1974.

119. "Urban Ozone and the Clean Air Act: Problems and Proposals for Change," Staff paper, U.S. Office of Technology Assessment, April 1988.

120. H. Buchner, "Hydrogen Use—Transportation Fuel," *International Journal of Hydrogen Energy*, Vol. 9, pp. 501–514, 1984; H. Buchner and R. Povel, "The Daimler-Benz Hydride Vehicle Project," *International Journal of Hydrogen Energy*, Vol. 7, pp. 259–266, 1982; "Hydrogen Fuel: Problems and Promises," *Automotive Engineering*, Vol. 88, No. 1, pp. 42–45, January 1980; R.E. Billings, "A Hydrogen Powered Mass Transit System," *International Journal of Hydrogen Energy*, Vol. 3, p. 49, 1978; M.A. DeLuchi, "Hydrogen Vehicles: An Evaluation of Fuel Storage, Performance, Safety, Enviromental Impacts, and Cost," *International Journal of Hydrogen Energy*, Vol. 14, No. 2, pp. 81–130, 1989.

121. About 500,000 natural gas-powered cars are in use worldwide. See, e.g: R.N. Abram, A.L. Titchener, and J.P. West, "Report on Overseas Visit to Investigate Compressed Natural Gas in Italy," New Zealand Liquid Fuels Trust Board, November 1979; R.L. Bechtold, T.J. Timbario, R.R. Tison and R.J. Sprafka, "The Practical and Economic Considerations of Converting Highway Vehicles to Use Natural Gas as a Fuel," *Proceedings of the Society of Automotive Engineers Conference P-129*, Pittsburgh, PA, pp. 47–69, June 22–23, 1983.

122. Even when compressed to 2400 psi, hydrogen gas contains only 5% the energy per unit volume of gasoline. The energy per unit volume for metal hydrides is about 15% that of gasoline, for liquid hydrogen about 26% (Table 14).

123. Estimates of cargo space from EPA *1986 Mileage Guide*, DOE/CE-0019/4, October 1985.

124. The energy in a 20 gallon tank of gasoline (LHV) is

$$20\ gal. \times 3.78\ liters/gal \times 32.3\ MJ/liter = 2.44\ GJ.$$

The volume energy density in GJ/m^3 (LHV) from Table 14 is:

Hydrogen gas (2400 psia) : 1.76
Natural gas (2400 psia) : 5.51
Advanced battery : 0.90
Lead-acid battery : 0.42

To store 2.44 GJ requires a volume (in cubic meters):

Hydrogen gas (2400 psia) : 1.39
Natural gas (2400 psia) : 0.44
Advanced battery : 2.71
Lead-acid battery : 5.81

*125. For a vehicle with a range of 200 miles (322 km), the storage volumes needed for various fuels and fuel economies are:

FUEL	Energy Density (GJ/m³)	Fuel volume (in m³) if fuel economy is:			
		12 mpg	25 mpg	50 mpg	100 mpg
Hydrogen gas @ 2400 psi	1.76	1.16	0.56	0.28	0.14
Metal hydride	4.9	0.42	0.20	0.10	0.050
Natural gas @ 2400 psi	5.51	0.37	0.18	0.089	0.044
Lead-acid Batteries	0.42	4.86	2.33	1.17	0.58
Advanced Batteries	0.90	2.27	1.09	0.54	0.27

126. Energy and Environmental Analysis, "The Motor Fuel Consumption Model, Twelfth Periodical Report," prepared for the Office of Policy, Planning, and Analysis, U.S. Department of Energy, DOE/PE/77000-1, November 1985.

127. J. Goldemberg, T.B. Johansson, A.K.N. Reddy, and R.H. Williams, *Energy for a Sustainable World, op. cit.* (Note 33).

128. Deborah Bleviss, *Preparing for the 1990s: the World Automotive Industry and Prospects for Future Fuel Economy Innovation in Light Vehicles,* Federation of American Scientists, Washington, DC, January 1987.

129. S.L. Fawcett and J.C. Swain, "Prospectus for a Consumer Demonstration of a 100 MPG Car," Battelle Memorial Institute Paper, March 1983.

130. Bleviss, *Preparing for the 1990s.*

131. R.R. Sekar and R. Kamo [Cummins Engine Company, Columbus, Indiana]

and J.C. Wood [NASA Lewis Research Center, Cleveland, Ohio], "Advanced Adiabatic Engines for Passenger Cars," Paper 840434, presented at the International Congress & Exposition of the Society of Automotive Engineers, Detroit, Michigan, February 27–March 2, 1984.

*132. The cost of liquefying, storing and dispensing liquid hydrogen as a transportation fuel is discussed in this note.

If one day's storage is desired, the cost of liquefaction is shown in the box at the bottom of this page.

See J.B. Taylor, J.E.A. Alderson, K.M. Kalyanam, A.B. Lyle, and L.A. Phillips, "Technical and Economic Assessment of Methods for the Storage of Large Quantities of Hydrogen," *International Journal of Hydrogen Energy,* Vol. 11, pp. 5–22, 1986.

133. The losses for transferring small quantities of liquid hydrogen can be as much as

Liquefaction Rate (tonnes/day)	Storage Capacity (tonnes)	Capital Cost ($/GJ)	Power Cost ($/GJ)	O&M Cost ($/GJ)	Total Liquefaction Cost ($/GJ)
30	30	4.10	4.34	1.77	10.20
100	100	2.14	4.34	0.92	7.40
300	300	1.28	4.34	0.55	6.17

50%, when a cold gas recovery loop is not included. See J.B. Taylor, J.E.A. Alderson, K.M. Kalyanam, A.B. Lyle, and L.A. Phillips, "Technical and Economic Assessment of Methods for the Storage of Large Quantities of Hydrogen," *International Journal of Hydrogen Energy*, Vol. 11, pp. 5–22, 1986; M.A. DeLuchi, "Hydrogen Vehicles: An Evaluation of Fuel Storage, Performance, Safety, Enviromental Impacts, and Cost," *International Journal of Hydrogen Energy*, Vol. 14, No. 2, pp. 81–130, 1989.

*134. The volume needed to store the energy equivalent of 2 gallons of gasoline as gaseous hydrogen at 2400 psia is 139 liters.

The weight of the needed containers can be estimated for pressurized gas cylinders as follows:

Cylinder type (rated 2400 psia)	volume (liters)	weight (kg)	installed cost (1986$)
Steel	50	52	281
Aluminum	31	26	253

For the example above, the added storage weight is 156 kg for 3 steel cylinders or 130 kg for 5 aluminum cylinders.

The installed cost for 5 aluminum cylinders would be $1265. For three steel cylinders it would be $843.

See R.N. Abram, A.L. Titchener, and J.P. West, "Report on Overseas Visit to Investigate Compressed Natural Gas in Italy," New Zealand Liquid Fuels Trust Board, November 1979; R.L. Bechtold, T.J. Timbario, R.R. Tison and R.J. Sprafka, "The Practical and Economic Considerations of Converting Highway Vehicles to Use Natural Gas as a Fuel," Society of Automotive Engineers Confer-

ence P-129, Pittsburgh, PA, June 22–23, 1983, pp. 47–69.

*135. Filling station costs for gaseous fuels are estimated in this note. In all cases it is assumed that natural gas or hydrogen is available to the filling station at a local pipeline pressure of 300 psia.

For the cases where compressed gas (natural gas or hydrogen) is delivered to onboard gas cylinders in cars, the cylinder pressure is assumed to be 2400 psia, and the gas is assumed to be transferred during refueling from intermediate storage tanks or "cascades" at the filling station maintained at a pressure of 3600 psia. Refuelling takes about 6 minutes with this system.

For the case of hydride storage, hydrogen is compressed from pipeline pressure to about 50 atmospheres (735 psia) and piped directly into the hydride tank. Refuelling takes about 10 minutes per car.

The fuel cost increment for the filling station ($ per GJ) is estimated to be:

VEHICLE STORAGE TYPE:		
Compressed Natural Gas	Compressed Hydrogen	Metal Hydride
1.57	2.94	1.45

The lower pressures needed to recharge hydride tanks make this option less expensive than compressed hydrogen gas.

See R.N. Abram, A.L. Titchener, and J.P. West, "Report on Overseas Visit to Investigate Compressed Natural Gas in Italy," New Zealand Liquid Fuels Trust Board, November 1979; R.L. Bechtold, T.J. Timbario, R.R. Tison and R.J. Sprafka, "The Practical and Economic Considerations of Converting Highway

Vehicles to Use Natural Gas as a Fuel,'' Society of Automotive Engineers Conference P–129, Pittsburgh, PA, June 22–23, 1983, pp. 47–69; A. Golovoy and R.J. Nichols, ''Natural Gas Powered Vehicles,'' *ChemTech*, June 1983, pp. 359–363.

*136. Hydride storage would weigh about as much as gaseous storage, would take up about 1/3 as much space, and would cost about half to two thirds as much.

See H. Buchner, ''Hydrogen Use—Transportation Fuel,'' *International Journal of Hydrogen Energy*, v. 9, 1984, pp. 501–514; H. Buchner and R. Povel, ''The Daimler-Benz Hydride Vehicle Project,'' *International Journal of Hydrogen Energy*, v. 7, 1982, pp. 259–266.

137. In this system, a liquid hydride (methylcyclohexane, for example) is broken down (using heat input from the engine exhaust) in an onboard ''rehydrogenation'' unit to form hydrogen and toluene. The hydrogen is burned in the vehicle's engine, and the toluene is stored in a tank, for subsequent recharging with hydrogen. The potential advantage of this system is that methylcyclohexane and toluene are liquids at standard temperature and pressure, which could reduce local distribution and filling station costs. (See T.H. Schucan, ''Seasonal Storage of Hydrogen and Use as a Fuel for Heavy Vehicles,'' Swiss Federal Institute for Reactor Research Report, October 1987.)

138. H. Buchner, *op.cit.*

139. John B. Heywood, ''Automotive Engines and Fuels: A Review of Future Options,'' Progress in Combustion Science, v. 7, 1981, pp. 155–184.

140. Spark-ignited direct-injection (or stratified charge) engines run equally well on a variety of fuels ranging from gasoline to Diesel fuel. These engines could also be used with hydrogen. The efficiency may be as much as 50% higher than that of a conventional spark-ignited gasoline engine. See J.M. Lewis and W.T. Tierney, ''Stratified Charge Engine Development with Broad Fuel Tolerance,'' First International Automotive Fuel Economy Research Conference, Arlington, Virginia, October 31–November 1, 1979; E. Mitchell and M. Alperstein, ''Texaco Controlled-Combustion System—Multifuel, Efficient, Clean and Practical,'' Combustion Science and Technology, v. 8, 1973, pp. 39–49; T. Krepec, T. Tebelis, C. Kwok, ''Fuel Control Systems for Hydrogen Fuelled Automotive Combustion Engines,'' *International Journal of Hydrogen Energy*, v. 9, 1984, pp. 109–114; P.C.T. de Boer, W.J. McLean, and H.S. Homan, ''Performance and Emissions of Hydrogen Fuelled Internal Combustion Engines,'' *International Journal of Hydrogen Energy*, v. 1, 1976, pp. 153–172.

141. *See Note 135.*

*142. In this note the cost of owning and operating a car is estimated for various fuels (gasoline, synthetic natural gas, methanol, ethanol, hydrogen and batteries) and various levels of automotive fuel economy. The results are shown in Figure 14.

143. In general, the larger the vehicle, the less important the weight and volume of the fuel storage system. For a hydrogen-powered subcompact car weighing 800 kg, with fuel stored in four 30-liter aluminum pressurized gas cylinders, which weigh 26 kg each, about 12% of the vehicle weight will be due to the storage system. For a 4000 kg truck with a comparable range, the fuel storage system would account for only about 7% of the total vehicle weight. The relatively large storage volume needed for hydrogen would also impose less of a constraint for trucks and buses than for automobiles.

144. Billings Energy Corporation of Provo, Utah, has built and operated several hydrogen-powered buses. See R.E. Billings, "A Hydrogen-Powered Mass Transit System," *International Journal of Hydrogen Energy*, Vol. 3, p. 49, 1978.

145. Frank von Hippel, "US Transportation Energy Demand," PU/CEES Report No. 111, Center for Energy and Environmental Studies, Princeton University, Princeton, New Jersey, 1981.

146. Using liquid hydrogen rather than jet fuel would reduce the fully loaded takeoff weight of a jumbo jet (Boeing 747 type) by perhaps 25%, with a fuel savings of about 12%. Hydrogen could be piped to the airport via a gas pipeline, and liquefied onsite. However, the cost of liquefaction and storage of liquid hydrogen is large [$6–$10/GJ *(see Note 132)*] even when large quantities of fuel are involved. When these costs are added to a delivered pipeline cost of perhaps $12-$16/GJ, the total cost of LH2 is $18-$26/GJ, as compared to the projected 2000 jet fuel price of $7.75/GJ. The 12% energy savings is outweighed economically by the extra fuel cost.

   See J.B. Taylor, J.E.A. Alderson, K.M. Kalyanam, A.B. Lyle, and L.A. Phillips, "Technical and Economic Assessment of Methods for the Storage of Large Quantities of Hydrogen," *International Journal of Hydrogen Energy*, Vol. 11, pp. 5–22, 1986.

*147. The CO_2 emissions from the production and utilization of various fuels are estimated in this note and shown in Table 17.

148. Office of Policy, Planning, and Evaluation, United States Environmental Protection Agency, *Policy Options for Stabilizing Global Climate*, draft report to Congress, February 1989; P.A. Okken and T. Kram (Energy Study Centre, Netherlands Energy Research Foundation, The Netherlands), "CH_4/CO-Emissions from Fossil Fuels Global Warming Potential," paper prepared for the IEA/ETSAP Workshop, Paris, June 1989; Appendix B in M.A. Deluchi, R.A. Johnston, D. Sperling, "Transportation Fuels and the Greenhouse Effect," UER-180, University Research Group, University of California, Berkeley, December 1987.

149. J. Pangborn and M.I. Scott, "Domestic Uses of Hydrogen," in *Hydrogen: Its Technology and Implications*, David A. Mathis, ed., Energy Technology Review No. 9, Noyes Data Corporation, 1976.

*150. A number of catalysts have been tried experimentally for catalytic hydrogen combustion. Researchers in the U.S. (J.B. Pangborn and J.C. Sharer at Institute of Gas Technology, Chicago, Ill.) have reported that platinum-coated anodized aluminum gave the best results for heating applications.

   The output of a room-sized catalytic space heater would lie in the range 5,000-10,000 BTU/h.

   For a 10,000 BTU/h heater, about 0.00364-0.00936 ounces of platinum costing $1.8-$4.7 would be required.

   Platinum raw material would not be a significant cost in the construction of the heater. Nor would platinum supply be a serious constraint. Even at a manufacturing level of ten million 10,000 BTU/h space heating units per year, only 2270 pounds of platinum would be required. The yearly platinum production in the U.S. in 1983 was 375,000 pounds.

   J.B. Pangborn, "Catalytic Combustion of Hydrogen in Model Appliances," 15th Intersociety Conference on Energy Conversion, Seattle, April 18–22, 1980, p. 1731; *1985 Statistical Abstract of the United States*.

Hydrogen catalytic heaters using stainless steel pads have also been built and tested by R.E. Billings. (See R.E. Billings, "Hydrogen Appliances," 7th World Hydrogen Energy Conference, Moscow, USSR, September 1988.)

151. The "Platinum Cat" is a natural gas-fuelled catalytic space heater for residential use, which consumes about 5200 BTU of natural gas per hour. Depending on the venting system, 80–90% of the heating value of the fuel (or 4160–4680 BTU/h) is delivered to the room. The heater costs about $350, plus an average of $100 for installation.

Source: B. Wells, Thermal Systems, Inc., Tumwater, Washington, private communications, 1987.

*152. Hydrogen produces some NO_x emissions, even with the relatively low temperatures of catalytic combustion.

The NO_x emissions from model catalytic hydrogen appliances have been measured in the range of $0.22-0.62 \times 10^{-4}$ lb/MBTU, well below the maximum permissible levels. Thus, the catalytic combustion of hydrogen should not pose an indoor air pollution problem.

Source: J. C. Sharer and J.B. Pangborn, "Utilization of Hydrogen as an Appliance Fuel," in *Hydrogen Energy: Part B*, T.N. Veziroglu, ed., Plenum Press, 1974.

*153. When combustion products from a hydrogen catalytic heater are vented directly into a room, high humidity due to water vapor buildup is a consideration. With a hydrogen heater vented directly to the room, we find that the humidity would be 35–45% for both conventional and super-insulated houses.

The generally accepted indoor comfort range is a relative humidity of 20–50%,

and a temperature of 20–28 °C. The added humidity from the hydrogen heater could be beneficial on very cold days.

One might expect some condensation to occur in conventional houses on cold surfaces, such as single-glazed windows, on the coldest days. For superinsulated houses with double or triple glazing, the inner surface of the window should be warm enough to avoid this problem.

Sources: ASHRAE Handbook, 1981 Fundamentals, American Society of Heating, Refrigeration and Air Conditioning Engineers, Atlanta, Georgia, 1981.

J. Mercea, E. Grecu, and T. Fodor, "Heating of a Testing Room by Use of a Hydrogen-Fuelled Catalytic Heater," *International Journal of Hydrogen Energy,* vol. 6, No. 4, pp. 389–395, 1981.

154. J.B. Pangborn and J.C. Sharer, "Catalytic Fluid Heater," US Patent No. 3955556, May 11, 1976.

R.J. Dufour, "Heat Exchange Apparatus," US Patent No. 3916869, November 4, 1975.

155. M. Haruta, Y. Souma, and H. Sano, "Catalytic Combustion of Hydrogen-II," *International Journal of Hydrogen Energy,* v. 7, No. 9, pp. 729–736, 1982.

M. Haruta and H. Sano, "Catalytic Combustion of Hydrogen-III," *International Journal of Hydrogen Energy,* v. 7, No. 9, pp. 737–740, 1982.

M. Haruta and H. Sano, "Catalytic Combustion of Hydrogen-IV," *International Journal of Hydrogen Energy,* v. 7, No. 10, pp. 801–807, 1982.

156. Catalytic space heaters include a case, controls for starting combustion and

adjusting the heat output, and the catalyst material. In addition the cost of fabricating the heater must be considered. Based on conversations with catalyst manufacturers, a catalyst/heating element cost of about $80-$100/square foot seems reasonable for natural gas or hydrogen heaters.

The "Platinum Cat" natural gas heater (*see note 151*) contains about 1.3 square feet of heating area and delivers 4420 BTU/h to the room (at 85% efficiency). The total cost of the heater is $350. If 1.3 × 100 = $130 of the cost is due to the heating surface, then the case, controls, and fabrication would account for perhaps $220.

A hydrogen catalytic space heater designed by Dufour and tested by Pangborn and Sharer delivers 7000-9000 BTU/h/sq.ft. of heating surface area. For a 97% efficient hydrogen heater designed to deliver 4420 BTU/h, the required heating area would be about 0.55 square feet. The cost of the heating element would be about $55. Assuming that the controls, case and manufacturing costs are the same, the hydrogen heater would cost about $220 + 55 = $275.

For new construction, the installation cost for a natural gas catalytic heater is estimated to be about $100 per 5200 BTU/h unit. This includes running gas and electric connections to the heater and building a vent to the outside. Building the vent is by far the most time consuming and expensive step, so that the installation cost for an unvented hydrogen heater cost would probably be much less than $100.

Sources: B. Wells, Thermal Systems, Inc., Tumwater, Washington, private communications, 1987; representatives of the Jaeger Products Company, private communications, 1987; R.J. Dufour, "Heat Exchange Apparatus," US Patent No. 3916869, November 4, 1975, J.B. Pangborn, "Catalytic Combustion of Hydrogen in Model Appliances," 15th Intersociety Conference on Energy Conversion, Seattle, April 18-22, 1980, p. 1731.

*157. In a conventional cylindrical natural gas tank-type heater, the flue gases from an open flame burner at the bottom of the tank pass through a central pipe, heating the surrounding water, before being vented to the outside. The thermal efficiency is about 50-55%. A hydrogen water heater with essentially the same design, cost and performance could be constructed by using a burner with different-sized holes.

Catalytic hydrogen water heaters have been designed that are similar to natural gas fueled tank-type storage heaters. In a catalytic hydrogen heater, the central pipe is replaced by a pair of cylindrical catalytic surfaces which heat the surrounding water. If the steam from combustion is vented to the room, the efficiency of water heating would be 84%. If the steam from combustion is condensed to preheat incoming water, the efficiency of water heating would be close to 100%.

A tank-type water heater is made up of the tank, controls, and the heating elements. The cost of the water storage tank and controls should be about the same for a conventional design heater and for the catalytic heater. The differences in cost will arise from the cost of the heating elements. In addition, the conventional water heater will require a vent costing perhaps $50, while the catalytic hydrogen heater will not.

In the conventional tank-type heater, the burner and central pipe are assumed to contribute about half of the bare unit cost. Installation costs are assumed to be approximately equal to the bare unit cost. Thus, the heating elements are assumed to contribute 25% of the total installed cost of the heater.

The installed cost of the heater can be estimated as follows:

	H$_2$ Catalytic	H$_2$ or Nat. Gas Tank-Type
Open Flame Burner ($)	–	80
Catalytic Surface ($)	185	–
Rest of Tank ($)	240	240
Vents ($)	–	50
Total Installed Cost of Water Heater Plus Vents	428	370

Sources: "TAG Technical Assessment Guide, Volume 2: Electricity End Use, Part I: Residential Electricity Use," EPRI, P-4463-SR, September 1987, J.B. Pangborn, "Catalytic Combustion of Hydrogen in Model Appliances," 15th Intersociety Conference on Energy Conversion, Seattle, April 18–22, 1980, p. 1731.

*158. In this note the heat demand for space heating with alternative levels of the house's thermal integrity (conventional and "superinsulated" house) and the demand for domestic hot water are estimated.

When designing a space heating system for a house, the heating load on the coldest expected day of the year ("design day") must be estimated.

Heat losses in houses are primarily due to transmission and infiltration. Using standard engineering formulas, these losses can be computed.

Heat gains in a house come from the sun, people, lights and appliances. For a household with 3.06 people and the most energy-efficient appliances available in 1982, the heat gains were estimated to be 53,000 BTU/day, or on average 2208 BTU/h (M.H. Ross and R.H. Williams, *Our Energy: Regaining Control*, McGraw-Hill, 1981). If average efficiency appliances were used, the heat gain was found to be 77,000 BTU/day or 3208 BTU/h.

Assuming three people occupy the house, on average, and:

A_f = Floor area = 1500 ft^2
V = Volume of house = 12,000 ft^3
Width = 30 ft
Length = 50 ft
T_{in} = Indoor Temperature = 70°F
T design = Coldest Expected Outdoor Temp = 14°F (for Newark, NJ or New York City)

The maximum heat delivery rate for the space heating system was estimated to be:

	Conventional	Superinsulated
Heat Transmission Loss (BTU/h)	21,492	8362
Infiltration Heat Loss (BTU/h)	9677	3084
Heat Gain (BTU/h)	3208	2208
	1980 average appliances	1982 efficient appliances

117

	Conventional	Superinsulated
Heat Required from Heating System (BTU/h)	27961	9780

(This is the heat which must be delivered to the conditioned space. The fuel use will depend on the efficiency of the system.)

ESTIMATES OF ANNUAL ENERGY DEMAND

The annual space heating energy demand can be estimated, if the house design and weather data are known. We have used formulas from the EPRI Technical Assessment Guide for the annual energy consumption in the New York City area. We estimated:

	Conventional	Superinsulated
Annual Space Heating Energy Demand (MBTU/yr)	45.2	5.32

The annual water heating energy demand can be estimated, assuming each person uses 15.6 gallons/day of hot water at 120°F, the energy needed for water heating would be 9.27 MBTU/yr. (This is the energy which must be delivered to the water. The fuel use will depend on the efficiency of the water heater.)

Sources: "TAG Technical Assessment Guide, Volume 2: Electricity End Use, Part I: Residential Electricity Use," EPRI, P-4463-SR, September 1987; R.H. Wiliams, 'Policy Proposals to Promote Energy Conservation in the Residential Sector in New Jersey," 1983; ASHRAE Handbook, 1981 Fundamentals, American Society of Heating, Refrigeration and Air Conditioning Engineers, Atlanta, Georgia, 1981.

*159. In this note the cost and performance of space heating and water heating systems are estimated for three different heating fuels, natural gas, hydrogen and electricity and three sets of energy end-use technologies:

1. *Low First-Cost Technologies for Conventional New Houses:* conventional furnaces for natural gas and hydrogen; conventional tank-type water heaters for natural gas and hydrogen; electric resistance space and water heating for electricity.

2. *Energy-Efficient Technologies for Conventional New Houses:* condensing furnaces for natural gas and hydrogen; tankless water heater for natural gas; catalytic water heater for hydrogen; and heat pumps for space and water heating with electricity.

3. *Energy-Efficient Technologies for Superinsulated Houses:* catalytic space heaters for both natural gas and hydrogen; tank type water heater for natural gas, catalytic water heater for hydrogen, resistance space heating and heat pump water heater for electricity.

The heat demands for each case are taken from Note 158 for conventional and super-insulated houses in New Jersey.

The heating unit for each case can be sized by dividing the peak heat demand by the heating unit's efficiency. The performance, size and costs for hydrogen space and water heaters were estimated as in Notes 156 and 157.

Cost and performance figures for the various systems are summarized in Table 20.

*160. Using the information in Tables 19 and 20, the levelized annual cost of heating can be calculated for each case. We have plotted the levelized annual cost of space and water heating for conventional and super-insulated houses for natural gas, hydrogen and electricity in Figure 16.

161. One example of a superinsulated house is the Northern Energy Home (NEH) offered in New England. It is difficult to estimate the added cost of a superinsulated house such as the NEH because the modular construction of the NEH facilitates both a high level of quality control and rapid site construction. As a consequence, the first costs of these houses have not been notably higher than conventional with more typical heat loss characteristics (see R.H. Williams, G.S. Dutt, and H.S. Geller, "Future Energy Savings in US Housing," *Annual Review of Energy*, Vol. 8, pp. 269–332, 1983.) However, in Sweden, where most new houses are prefabricated, one manufacturer (Faluhus) has offered both conventional and superinsulated versions of the same basic prefabricated housing design: the superinsulated design costing $3140 more requires 32.3 GJ less heating energy per year than the conventional version (J. Goldemberg, T.B. Johansson, A.K.N. Reddy, R.H. Williams, *Energy for a Sustainable World*, World Resources Institute, Washington DC, September 1987). From Note 158, our superinsulated house uses 5.32 MBTU/yr (5.61 GJ/yr) for space heating as compared to 45.2 MBTU/yr (47.7 GJ/yr) for the conventional house, resulting in a savings of 42.1 GJ/year. Assuming the initial cost is proportional to the savings the Faluhus experience suggests that the superinsulated house considered here, which requires 42.1 GJ less heat per year than a house of conventional construction would cost $4100 more to build than a similar house equiped with conventional amounts of insulation. This estimate is consistent with an EPRI estimate that superinsulated houses (in the 1500 square foot size) would cost $3764–$6428 more than conventional houses (Electric Power Research Institute, *Technical Assessment Guide, Volume 2: Electricity End Use, Part I: Residential Electricity Use*, EPRI, P-4463-SR, September 1987).

*162. In this note the value of the credit for the oxygen byproduct is estimated to be $1.5–$2.2 per GJ of hydrogen produced, depending on the pressure to which the oxygen is compressed.

Sources: M. Hammerli, *International Journal of Hydrogen Energy*, v. 9, No. 12, pp. 25–51, 1984; B. Kribel, Air Products, Inc., private communications, 1987.

163. W. Balthasar, "Hydrogen Production and Technology: Today, Tomorrow and Beyond," *International Journal of Hydrogen Energy*, v. 9, pp. 649–668, 1984.

*164. Here the cost of industrial hydrogen is estimated for alternative production methods.

165. E. Fein and K. Edwards, "Market Potential of Electrolytic Hydrogen Production in Three Northeastern Utilities' Service Territories," EPRI Report, EM-3561, May 1984.

166. W. Balthasar, *op. cit.*

167. Energy Information Administration, *Annual Energy Outlook 1986, with Projections to 2000*, DOE/EIA-0383(86), February 1987.

168. P.K. Takahashi and S.H. Browne, "Hydrogen Energy from Renewable Sources," Hawaii Natural Energy Institute report, 1987.

169. W. Balthasar, *op. cit.*

170. Large electrolysis plants sited near hydro-electric power are:

Location	Type	Manufacturer	Year Installed	Power In (MW)	H_2 Out (10^6 scf/hr)
Trail, BC	Unipolar	Cominco	1939	n.a.	0.7
Nangal, India	Bipolar	DeNora	1958	130	1.1
Aswan, Egypt	Bipolar	DeMag	1960–1977	170	1.4
Rjukan, Norway	Bipolar	Norsk	1927–1965	250	2.2
Cuzco, Peru	Bipolar	Lurgi	1958	25	0.2

Source: E. Fein and T. Munson, "An Assessment of Non-Fossil Hydrogen," Gas Research Institute Report, GRI 79/0108, 1980.

171. E. Fein and K. Edwards, *op. cit.*

172. K. Darrow, N Biederman, and A. Konopka, "Commodity Hydrogen from Off-Peak Electricity," *International Journal of Hydrogen Energy*, v. 2, pp. 175–187, 1977; R.W. Foster, R.R. Tison, W.J.D. Escher, and J.A. Hanson, "Solar/Hydrogen Systems Assessment," DOE/JPL-955492, 1980.

173. E. Fein and K. Edwards, *op. cit.*

174. J. Nitsch and C.-J. Winter, "Solar Hydrogen Energy in the F.R. of Germany: 12 Theses," *International Journal of Hydrogen Energy*, v. 12, p. 663, 1987.

175. *See Note 164.*

176. Maricopa County Regional Public Transit Authority, "Building Mobility: Transit 2020," draft report, Phoenix, Arizona, 1988.

177. *Ibid.*

178. A.J. Pfister, General Manager, Salt River Project, Phoenix, Arizona, private communications, 1988.

*179. Suppose that 30,000 fleet automobiles driven 100 miles per day, on average, and having an average fuel economy of 30 mpg (gasoline equivalent) are refuelled every night with hydrogen (metal hydride storage) derived via electrolysis using off-peak AC electricity from local conventional power sources.

The required fuel would be 100,000 gasoline-equivalent gallons per day, or 12,500 GJ per day (386 MW) of hydrogen energy. Assuming an electrolyzer efficiency of 80.6% and off-peak power available 9 hours a day (10pm to 7am), the required off-peak electric power is 480 MW.

As there are no scale economies for hydrogen production above 10 MW, there could be 39 hydrogen charging stations in Phoenix, each servicing 769 cars per night.

The cost of hydrogen for these cars would be as follows (in $/GJ):

Electricity for Electrolyzers	6.89 to 10.33
Electrolyzer	2.34
Capital Charge for Compressors	1.01
Electricity Cost for Compressors	0.34 to 0.50
Capital Charge for Storage	0.13
Capital Charge for Refuelling Bays	0.020
Labor Cost for Operating Refuelling Bays	0.79
Administrative, Overhead Costs for Refuelling	0.44
Total	11.96 to 15.69

180. M.C. Holcomb, S.D. Floyd, S.L. Cagle, "Transportation Energy Data Book: Edition 9," Oak Ridge National Laboratory, ORNL-6325, April 1987.

181. *See Note 142.*

*182. The estimated fixed capital required to convert 30,000 fleet vehicles to hydrogen produced off peak using existing power sources is the following, in million dollars:

Electrolyzers	101.5
Compressors	22.9
Hydrogen Storage	6.2
Refuelling Bays	0.9
Total	131.5

183. It is assumed that cars last 105,000 miles. Driven 100 miles a day for 350 days per year, they thus last 3 years. In contrast, the hydrogen system is expected to last 20 years, and serve many generations of fleets. Assuming straight line depreciation and a 6.1% discount rate, the capital cost allocated to the first generation of fleets is:

$131.5 \times 10^6 \times [1 - (17/20)/1.061^3]/30,000 =$ $1264 per car.

184. Metal hydride tanks are estimated to cost $635 (*see Note 142*) and last more than 2000 cycles. But in 3 years of operation in fleet cars they would use only 1050 cycles, if refuelled every night for 350 days per year. Thus it is assumed that hydride tanks are used in two generations of fleet vehicles. Assuming straight line depreciation, the cost allocated to the first generation of vehicles is:

$635 \times [1 - 0.5/1.061^3] = $369 per car.

185. Maricopa County Regional Public Transit Authority, "Building Mobility: Transit 2020."

*186. Suppose that by the year 2000, 30,000 additional fleet automobiles driven 100 miles per day, on average, and having an average fuel economy of 30 mpg (gasoline equivalent) are refuelled every night with hydrogen (metal hydride storage) derived via electrolysis using DC electricity a PV electric system.

The required fuel would be 100,000 gasoline-equivalent gallons per day, or 12,500 GJ per day of hydrogen energy. Assuming an electrolyzer efficiency of 84% the electricity requirement would be some 1.45 billion kWh per year. For insolation similar to El Paso, the electricity production rate is 2.37 kWh per year per peak Watt, so that the required installed electric capacity is 610 MW.

As there are no scale economies for PV hydrogen production above 10 MW, there could be 39 hydrogen charging stations in Phoenix, each servicing 769 cars per night.

The cost of hydrogen for these cars would be as follows (in $/GJ):

Electricity for Electrolyzers	6.57 to 11.47
Electrolyzers	2.52
Capital Charge for Compressors	1.01
Electricity Cost for Compressors	0.30 to 0.51
Capital Charge for Storage	1.18
Capital Charge for Refuelling Bays	0.020
Labor Cost for Operating Refuelling Bays	0.79
Administrative, Overhead Costs for Refuelling	0.44
Total	12.83 to 17.94

187. M.C. Holcomb, S.D. Floyd, S.L. Cagle, "Transportation Energy Data Book: Edition 9."

188. The cost of pipeline transmission depends on the pipeline flow rate. There is considerable economy of scale for pipelines up to levels of perhaps 100,000 million BTU/hr. To supply this flow rate, an electrolyzer would have to have a peak output of 29,000 MW of hydrogen. Still the overall contribution of transmission costs over 1000 miles is a small part of the total cost even for flow rates of 20,000 million BTU/hr (corresponding to an electrolyzer about 6000 MW in size).

Sources: K.E. Cox and K.D. Williamson, *Hydrogen: Its Technology and Implications,* CRC Press, Cleveland, Ohio, 1975; D. Christodoulou, "Technology and Economics of the Transmission of Gaseous Fuels," Princeton University Master's Thesis, 1984.

189. J.M. Hall, Sales Manager, RIX Industries, Oakland, CA, private communication, 1988.

190. "Fact Sheet: President Bush's Clean Air Plan," White House Press release, June 12, 1989.

191. H.G. Broadman and W.W. Hogan, "Oil Import Policy in an Uncertain Market," Energy and Environmental Policy Discussion Paper No. E-86-11, John F. Kennedy School of Government, Harvard University, November 1986.

192. W.U. Chandler, H.S. Geller, and M.R. Ledbetter, *Energy Efficiency: A New Agenda,* American Council for an Energy Efficient Economy, Washington, DC, July 1988.

193. J. Goldemberg, T.B. Johansson, A.K.N. Reddy, and R.H. Williams, *Energy for a Sustainable World,* See Note 33.

194. *Ibid.*

195. K. Zweibel, SERI, Golden, Colorado, private communication, 1989.

196. J.O'M. Bockris, Texas A&M University, private communication, 1989.

197. *See Note 174.*

198. As the era of low-cost oil and gas draws to a close, financing energy supply expansion will become more and more challenging, because the major alternative energy supplies are much more capital-intensive than in the past. The table below compares the unit capital costs required for alternative energy systems.

Bringing PV hydrogen production capacity on line would require capital expenditures of \$3.7-\$5.8/(Watt-yr/yr) for PV arrays and electrolyzers. *See Note 94.* (Hydrogen storage systems and pipelines would be a relatively small addition to this cost.)

Coal-based synthetic fuels would require capital expenditures of \$0.6-\$1.1/(Watt-yr/yr) for plants to produce synthetic natural gas, and \$1.0-\$2.5/(Watt-yr/yr) for plants producing synthetic liquids (methanol and synthetic gasoline). *See Note 114.* (In addition to the cost of building synfuel plants, modest additional capital would be needed for expansion of coal mining and transport systems.)

The unit capital costs for electricity systems are estimated to be in the range 7.7 to 10 Watt-yr/yr during the period 1980-2000 (a).

Thus capital requirements for a PV hydrogen energy system would be itermediate between those required for electricity systems in the period 1980-2000 and those required for coal synthetics systems.

Development of a global PV hydrogen system capable of supplying 5 Terawatt-yr/yr (or 1 kW-yr/yr for each of 5 billion people) would cost 18-30 trillion dollars.

For comparison the cumulative capital investment expected between 1980 and 2000 for oil, natural gas, coal and electric energy systems is in the range 14 to 20 trillion dollars (or 700 to 1000 billion dollars per year) (a).

Thus, if the estimated 1980–2000 capital requirement for global energy is taken as an indicator of the level of capital investment society will make on energy systems it is apparent that building a worldwide PV hydrogen energy system would take many decades.

ESTIMATED CAPITAL REQUIRED FOR THE WORLD'S ENERGY SYSTEMS, 1980–2000[a]

	Low Energy Growth Scenario					High Energy Growth Scenario				
	Oil	Nat. Gas	Coal	Elec.	Total	Oil	Nat. Gas	Coal	Elec.	Total
Capital Cost per Watt-yr/yr [$/(Watt-yr/yr)]	0.69	0.62	0.20	7.7	1.3	0.75	0.79	0.33	10.1	1.7
Cumulative Capital Cost (Trillion $)	2.82	1.36	0.61	8.9	13.7	3.23	1.89	1.25	13.3	20

ESTIMATED CAPITAL COSTS FOR PV HYDROGEN SYSTEMS[b]

	PV module eff. = 18% PV module cost = $0.2/Wp			PV module eff. = 12% PV module cost = $0.4/Wp		
Capital Cost [$/(Watt-yr/yr)]	3.67			5.85		
Energy Production (TW-yr/yr)	1	5	10	1	5	10
Cumulative Capital Cost (Trillion $)	3.7	18.3	36.7	5.8	29.3	58.5

ESTIMATED CAPITAL COSTS FOR SYNTHETIC FUELS FROM COAL[c]

	Methanol		Syn. Gasoline		SNG	
Capital Cost [$/(Watt-yr/yr)]	1.0–2.2		2.5		0.6–1.1	
Energy Production (TW-yr/yr)	1	5	1	5	1	5
Cumulative Capital Cost (Trillion $)	1.0–2.2	5–11	2.5	12.5	0.6–1.1	3.0–5.5

a. H. K. Schneider and W. Schulz, "Investment Requirements of the World Energy Industries 1980–2000," World Energy Conference, London, September 1987.

b. These estimates are for PV hydrogen systems in the Southwestern U.S., with balance of systems cost = $33/m², electrolyzer capital cost = $170/kWDCin, electrolyzer efficiency = 0.84, electrolyzer-PV coupling efficiency = 0.93, and average insolation = 271 Watts/m².

c. For coal synthetic fuel plants in the U.S. See Note 114 for details.

World Resources Institute

1709 New York Avenue, N.W.
Washington, D.C. 20006 U.S.A.

The World Resources Institute (WRI) is a policy research center created in late 1982 to help governments, international organizations, and private business address a fundamental question: How can societies meet basic human needs and nurture economic growth without undermining the natural resources and environmental integrity on which life, economic vitality, and international security depend?

Two dominant concerns influence WRI's choice of projects and other activities:

The destructive effects of poor resource management on economic development and the alleviation of poverty in developing countries; and

The new generation of globally important environmental and resource problems that threaten the economic and environmental interests of the United States and other industrial countries and that have not been addressed with authority in their laws

The Institute's current areas of policy research include tropical forests, biological diversity, sustainable agriculture, energy, climate change, atmospheric pollution, economic incentives for sustainable development, and resource and environmental information.

WRI's research is aimed at providing accurate information about global resources and population, identifying emerging issues and developing politically and economically workable proposals.

In developing countries WRI provides field services and technical program support for governments and non-governmental organizations trying to manage natural resources sustainably.

WRI's work is carried out by an interdisciplinary staff of scientists and experts augmented by a network of formal advisors, collaborators, and cooperating institutions in 50 countries.

WRI is funded by private foundations, United Nations and governmental agencies, corporations, and concerned individuals.